The Nature of Time

This book reviews and contrasts contemporary and historical perceptions of time from scientific and intuitive human points of view. Ancient and modern clocks, Augustinian ideas, the deterministic Newtonian universe, biological clocks, deep time, thermodynamics, quantum mechanics, and relativity all contribute to the perspective. The focus is on what can be inferred from established technologies and science as opposed to futuristic speculation.

Chapter 1 describes clocks, including the cesium atomic clocks establishing the current global time standard, a history of clock development, biological clocks, phylogenetic trees, radioactive dating, and astronomical methods to determine the age of the universe. Chapter 2 poses ancient questions about time not fully addressed by an understanding of the technical nature of clocks. An early summary of some of these questions as described by Augustine in the 3rd century CE is followed by a description of how Newton, 1300 years later, introduced a conception of time which provided some answers, such as the nature of an infinitesimally short present. Implications concerning the reality of events in the past, present, and future are also discussed. The Newtonian picture is contrasted with the intuitive human one and the possibilities of time travel and temporal recurrence are briefly discussed. Chapter 3 introduces the second law of thermodynamics and addresses how it is compatible with a time-reversible Newtonian description of a universe, even though it appears to define an "arrow of time." The nature of entropy and its relation to coarse graining and emergence play a central role in the discussion. Chapter 4 discusses ways in which quantum mechanics has altered the Newtonian perspective, accounting for various interpretations of the meaning of quantum mechanics with regard to time. Chapter 5 describes basic elements of special relativity and their implications for the nature of time. Time dilation and the fact that even the temporal order of events can be different as recorded in different frames of reference are described. The examples are chosen to avoid evocation of currently unattainable technologies. An afterword in Chapter 6 reviews questions raised by Augustine and summarizes how the development of science since then has addressed them.

This book was originally developed for an interdisciplinary seminar for beginning undergraduates at the University of Minnesota. It uses a small amount of algebra, mainly in supplementary appendices, and does not assume any prior knowledge of physics, chemistry, biology, or astronomy. In contrast to many semipopular books on time, it avoids speculation either about engineering (techno-optimism) or physical theory (strings, loop quantum gravity, black hole entropy). Instead, it takes a more grounded approach and describes what is currently known (and not known) to help both students and the general reader make better sense of time.

J. Woods Halley is a Professor at the University of Minnesota in Minneapolis. His research group studies electrochemical phenomena, including the origin of life, as well as low temperature phases of many-body systems, including superfluidity and superconductivity, using analytical theory and computer simulation. He was educated at the University of California, Berkeley and the Massachusetts Institute of Technology and is a Fellow of the American Physical Society. He has previously published books on the likelihood of extraterrestrial life and statistical mechanics.

The Nature of Time

J. Woods Halley

CRC Press is an imprint of the
Taylor & Francis Group, an **informa** business

First edition published 2023
by CRC Press
6000 Broken Sound Parkway NW, Suite 300, Boca Raton, FL 33487-2742

and by CRC Press
4 Park Square, Milton Park, Abingdon, Oxon, OX14 4RN

CRC Press is an imprint of Taylor & Francis Group, LLC

Library of Congress Cataloging-in-Publication Data

Names: Halley, J. Woods (James Woods), 1938- author.
Title: The nature of time / J. Woods Halley.
Description: First edition. | Boca Raton : CRC Press, 2023. | Includes bibliographical references and index.
Identifiers: LCCN 2022013273 (print) | LCCN 2022013274 (ebook) | ISBN 9780367478759 | ISBN 9780367477066 | ISBN 9781003037125
Subjects: LCSH: Time. | Time measurements.
Classification: LCC QB209 .H24 2023 (print) | LCC QB209 (ebook) | DDC 529/.7--dc23/eng20220820
LC record available at https://lccn.loc.gov/2022013273
LC ebook record available at https://lccn.loc.gov/2022013274

ISBN: 978-0-367-47875-9 (hbk)
ISBN: 978-0-367-47706-6 (pbk)
ISBN: 978-1-003-03712-5 (ebk)

DOI: 10.1201/9781003037125

Typeset in Nimbus Roman
by KnowledgeWorks Global Ltd.

Publisher's note: This book has been prepared from camera-ready copy provided by the authors.

Contents

Preface

In this short volume I have summarized and expanded materials which I have used over several decades in small undergraduate classes on the nature of time at the University of Minnesota. The material is designed to be accessible to people with a little knowledge of algebra and possibly a little physics. It is not intended primarily for aspirant professional or professional physicists, though they are of course invited to peruse and possibly criticize it. I originally began the project when I ran across W. Whitrow's excellent but dated book on the Natural Science of Time. The subject and approach seemed ideal for our 'freshman seminar' series of courses at the University of Minnesota but I could not find more a contemporary book at a similar level and similarly comprehensive.

The subject matter here covers many fields of scientific specialty as the subject requires. That puts burdens on both the author and the reader, since I am not professionally trained in molecular biology or astronomy, for example, and those disciplines are extremely important to the story here. For readers, I have summarized some key needed features of physics, biology cosmology and relativity in a series of appendices, where I occasionally use a little more algebra than I do in the main text.

The understanding of time is not complete and some issues arise on which there is little or no professional consensus. That is particularly true for the implications of quantum mechanics on the nature of time (Chapter 4) but also for the nature of memory and consciousness (Chapter 2) and to a lesser extent the implications of the second law of thermodynamics (Chapter 3). In those cases, I have tried to describe a consistent perspective on the issues which I prefer while also acknowledging and providing references to other points of view.

I have tried throughout to present a perspective which is consistent with lived human experience as well as with currently established scientific fact. You will not find much here on such topics as quantum loop gravity, string theory, entropy of black holes, dark matter or other exotic but unestablished speculations which could bear on the subject if they are eventually confirmed to be relevant to the real world as we experience it. It is amusing to think about such things but if one mixes them together with established fact in a book intended for an unprofessional audience, confusion between what is established and what is speculation is very likely.

It is gratifying that many students have reported a good experience in the course which led to this book, and I thank them all for their interest as well as their input in term papers and discussion. Some materials here are expansions on student work. I also thank Joe Hautman, friend, former research collaborator and talented artist, for reading the manuscript and offering several useful criticisms as well as for permission to use Figure 3.3.

Comments and corrections from readers are, of course, welcome. I particularly invite corrections from professionals in those fields for which I am not a specialist. Teachers that might want to use some of these materials or the entire book in a class are welcome to contact me for lecture slides, problem sets and simple simulation codes which might be helpful.

J Woods Halley
Physics Department University of Minnesota Minneapolis
2021

CHAPTER 1

Clocks: The Nature of Time Measurement

In physical science a first essential step in the direction of learning any subject is to find principles of numerical reckoning and practicable methods for measuring some quality connected with it. I often say that when you can measure what you are speaking about and express it in numbers you know something about it; but when you cannot measure it, when you cannot express it in numbers, your knowledge is of a meagre and unsatisfactory kind: it may be the beginning of knowledge, but you have scarcely, in your thoughts, advanced to the stage of science, whatever the matter may be. William Thomson, Popular Lectures and Addresses Volume 1 Macmillan, London (1891)

1.1 MODERN CLOCKS

We will be primarily interested in conceptual issues concerning the nature of time. However, for that purpose, we need a preliminary idea of what people mean by the word time in everyday discourse. One answer is that time is what clocks measure. There is a danger of circularity in that definition but we will begin by looking closely at what contemporary clocks actually do and then consider what that implies concerning assumptions about the nature of time.

The world standard for time measurements is called Coordinated Universal Time. (The acronym UTC refers to the French name.) UTC is determined by averaging the results of several clocks in the US, France and many other countries (about 70 laboratories in all) and is used for many purposes such as coordinating the internet and making the Global Positioning System (GPS) work. (Details concerning how the coordination takes place are both bureaucratically and technically complicated. For example, the GPS has a clock which is maintained by the US Department of the Navy and is coordinated with the UTC clocks so that the duration of a second in the two clocks is identical (to the greatest precision possible) but the absolute value of the GPS time differs currently (1/2020) by .08 seconds. Reference [2] summarizes

DOI: 10.1201/9781003037125-1

properties of some of the important clocks, all coordinated with the UTC.) All the clocks are based on the international definition of the second which is

"The second is the duration of 9 192 631 770 periods of the radiation corresponding to the transition between the two hyperfine levels of the ground state of the cesium 133 atom".

as decided by international agreement with the stipulation added in 1997 that the cesium atom should be at the absolute zero of temperature. (I will explain later what the 133 designation for the cesium atom means. For the impatient and more knowledgable, it designates the isotope of cesium used, but you will not need to fully understand that to understand the following description of how the clock works.)

I will address three issues concerning this: 1) How is a clock based on such a definition constructed? 2) What does it imply concerning assumptions about the nature of the natural world? and 3) What does it imply concerning assumptions about the nature of time?

I will briefly describe one of the clocks used in determining the UTC. Until 2014 it was the US contribution to the UTC, built and maintained by the National Institute of Science and Technology (NIST), a US federal agency based in Boulder, Colorado. The current version of the clock is called NIST-F2 and is very similar. A picture of NIST-F1 is in Figure 1.1 What it does is as follows (Figure 1.2). Cesium metal in solid form is vaporized in a vacuum container. The cesium vapor is trapped by several crossed laser beams and manipulated to reduce its temperature to a very low value. Then the laser beams are used to put an upward force on the cesium atoms which tosses them up against the earth's gravitational field. The lasers are then turned off and the cesium atoms continue to move up about a meter and then turn around and fall down again like a ball tossed up in the air (but this is in a vacuum.) On the way up and on the way down, the cloud of cesium atoms passes through a microwave field

Figure 1.1 The cesium atomic clock NIST-F1 at the National Institute of Science and Technology in Boulder, Colorado. Reprinted courtesy of NIST, all rights reserved.

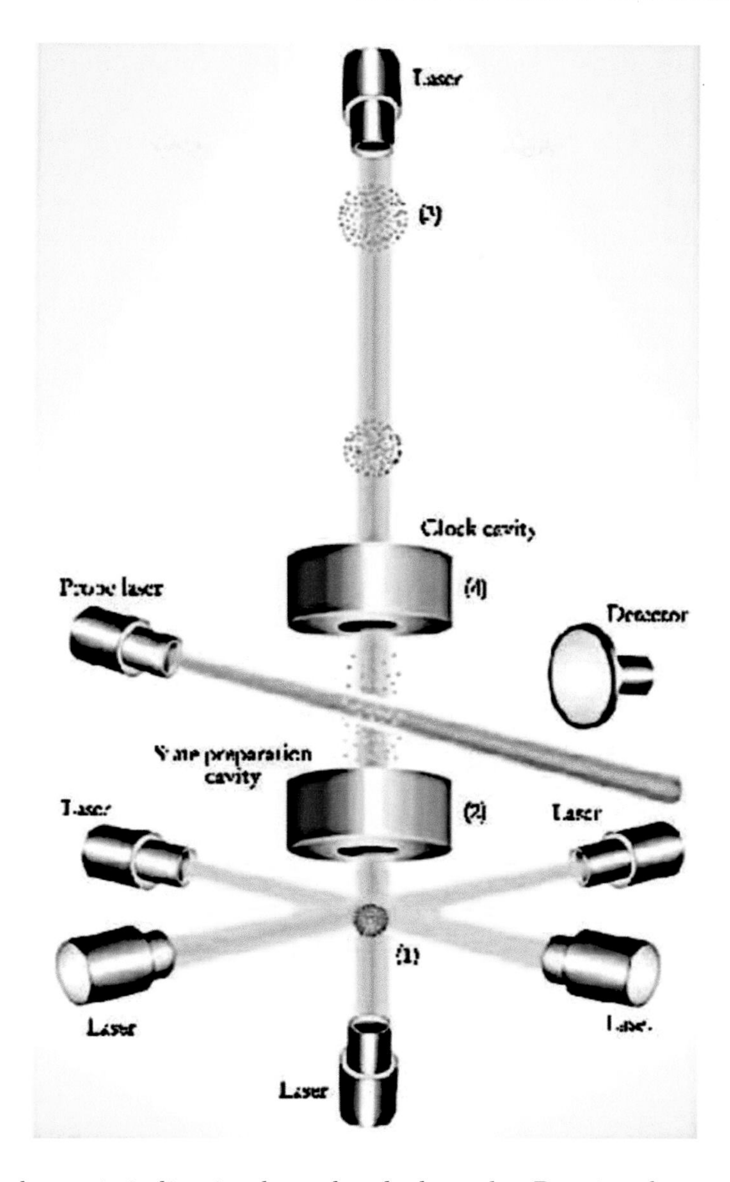

Figure 1.2 A schematic indicating how the clock works. Reprinted courtesy of NIST, all rights reserved.

whose frequency can be adjusted. When the atoms have passed back through the microwave field on the way down a probe laser is turned on which causes the cesium atoms which were perturbed by the microwave field to scatter visible light which is detected. Now the experiment is repeated over and over with various microwave frequencies until the highest intensity of visible scattered light is detected at the end. Then the microwave frequency is the same as the frequency of the cesium atom in the international definition of a second of time. The tuned microwaves are then used to drive a clock which is the standard clock used by GPS, the internet and so forth.

Some more details about what happens to the cesium atoms during this procedure are described in Appendix 1.1.
What does this clock assume about the nature of the world? First of all, it assumes that all cesium atoms containing nuclei of isotope 133 are exactly the same. There is a lot of evidence in favor of that assumption but it is impossible to ever prove it once and for all. The chemical identity of all atoms of a given chemical type is a relatively new idea (about two centuries old) in the history of science and the notion of isotopes is newer still (a little less than a century). Secondly, and maybe a little less obviously, one is assuming that the oscillations of the cesium atom are associated with the passage of something called time. At least that's one interpretation. Another interpretation is that the interval between the oscillations of the cesium atom DEFINES what we mean by time.
What does this clock assume about the nature of time? Well there is one such clock in Boulder, Colorado, a similar one in Paris, France, and other similar ones in 68 other laboratories around the world. It is assumed that they are all measuring the same time (time zones aside. Time zones just define the zero of time, not the length of a second or an hour.) This sounds like Newton's definition (or description) of time. In one translation from the Latin he wrote:

> Absolute, true, and mathematical time, from its own nature, passes equably without relation to anything external, and thus without reference to any change or way of measuring of time (e.g., the hour, day, month, or year).

(Some translations use the word 'flows' instead of 'passes'.) Notice that there is no mention of where the time is measured in this declaration. This statement of Newton's has been much discussed for the approximately three hundred years since it was published. Modern science does not look at time in quite that way, as we will discuss, but we will see that the contemporary way of measuring time conforms quite closely to it.

1.2 A BRIEF HISTORY OF THE MEASUREMENT OF TIME

Now that we've had a look at how time is measured (or defined) these days it will also be useful to review how we got here. That is because the history of the measurement of time can help reveal the origins of our intuitions and scientific assumptions about it. There are indications that time measurement of some sort has been carried out by humans for approximately 10,000 years. (The earliest human traces go back about 40,000 years, but we have no evidence of time measurement that far back.) The earliest measurements seem to have been based on astronomical observations of the regularity of the motions of objects in the sky particularly of the moon but also of the sun and the 'fixed stars'. One motivation for time measurement in early agricultural societies probably was to discern the appropriate time for planting crops. There is evidence that the power of religious elites in those early societies was based at least in part on their ability to read the skies correctly for that purpose. Relics of structures which can be shown to have had the astronomically related purpose of correctly timing the seasons survive from the Druids of Great Britain (Stonehenge) and the temples of the Mayan cultures of Central and North America. (Popular accounts of the Mayan calendar found on the web may not be reliable. For a professional source with references to the earlier scholarly literature, see [17].)

Though astronomical methods dominated, other methods of time measurement included the use of the rate of the human heart beat and the rhythm of music or poetry. Galileo, a pioneer in the development of physics who lived in the 17th century in Italy, used his pulse and a water clock for determining time intervals in his experiments. He also noted and used the regularity of a pendulum. Galileo is reported to have determined that the time between swings of a pendulum was a reliable indicator of the passage of time (even though the amplitude of the swings was getting smaller over time) by comparing the swinging rate to the rate of the beating of his heart.

It is illustrative that another scientist, Giovanni Battista Riccoli, who lived shortly after Galileo, mistrusted Galileo's results because he doubted the accuracy of the pendulum as a means of measuring time. So he made very careful comparisons of the rate of swinging of a pendulum with the time for the earth to make one full turn (a day). The story is told in detail in reference [25]. The basic idea was to compare time as measured by a pendulum to the time for a particular star to move a measured fraction of a day across the sky. (This is called sidereal time.) The method is illustrated in Figure 1.3. Riccoli's painstaking measurements established that the two times could be related to one another with great accuracy. When he used his carefully calibrated pendula in repetitions of Galileo's experiments on fallng bodies he confirmed Galileo's conclusions, contrary to his original expectation. The story is an exemplary one in terms of scientific practice. It also illustrates that the underlying assumption about time at that period of history, and long afterwards, was that, fundamentally, time was manifested in the regularity of motions of planets and particularly in the rate of rotation of the earth.

In fact, the length of the earth day remained the fundamental standard for the definition of the unit of time from the eighteenth century, when weights and measures were largely regularized through efforts in Europe after the French revolution, right up to the middle of the twentieth century, when atomic clocks were introduced.

Though Galileo and Riccoli used pendula as time keepers, those pendula were not really useful as clocks because they stopped swinging after a while due to the action of friction at the point from which the arm of the pendulum swung (as well as a small frictional effect due to the action of the air.) What was needed was a mechanical clock, using the regularity of pendulum, but providing a source of energy which would keep the pendulum going. The first known such mechanical pendulum clock was developed around 1635 by the astronomer Huygens. I provide a picture of Huygen's clock in Figure 1.4 and another diagram in Figure 1.5 which illustrates, in a simplified version, how such escapement clocks work.

The points to note are that there are two key elements to any mechanical clock. First, no mechanical mechanism will continue to function and indicate the time without some kind of source of organized mechanical energy. What I mean by 'organized' will be discussed in more detail later, particularly in Chapter 3, but here that source of energy is a weight, which is raised some distance above the surface of the earth

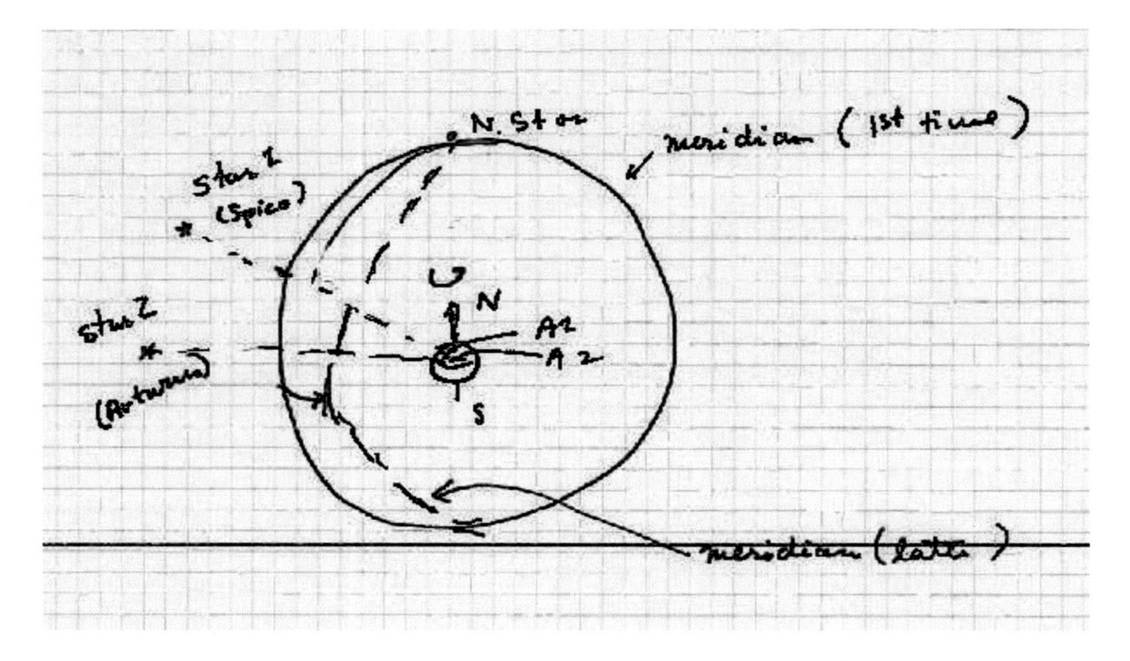

Figure 1.3 How Riccoli compared the rate that a pendulum swings to the rate at which the earth turns. The small sphere sketched at the center represents the earth, and the two points A1 and A2 are the positions of Riccoli and his helpers at two times during the night. They are different because the earth is turning, as indicated by the arrow. Riccoli (and any other observer of the sky in the northern hemisphere) can determine a line on the sky called the meridian by drawing a line through a point directly overhead through the north star and continuing from one horizon to the other. As the earth turns, that meridian moves across the sky with respect to the other stars. The meridian at the points when Riccoli et al are at A1 and (later in the night) at point A2 are sketched and labeled in Figure 1.3. Riccoli started his pendula swinging when the meridian crossed a star (recognized by its relationship to other stars in the sky) called Spica and counted the swings until the meridian crossed another easily visible star (Arcturus). One can work out the fraction of an earth day (technically a sidereal day) which separates the two crossings and thus determine how the period of the pendulum is related to the length of an earth day. Actually, Riccoli was more interested in whether the results of the pendulum swinging are reliably reproducible, so he did the experiment on three different nights. With the pendulum he was using, he got the same answer after more than 3000 swings to within 2 swings of the pendulum each time and concluded that the pendulum reliably measured time. With the particular pendulum he was using, each swing was taking one second, defined as $1/(24\times3600)$ of a sidereal day, within less than a percent. A sidereal day was defined as the time for a star to go once around and reappear at the meridian.

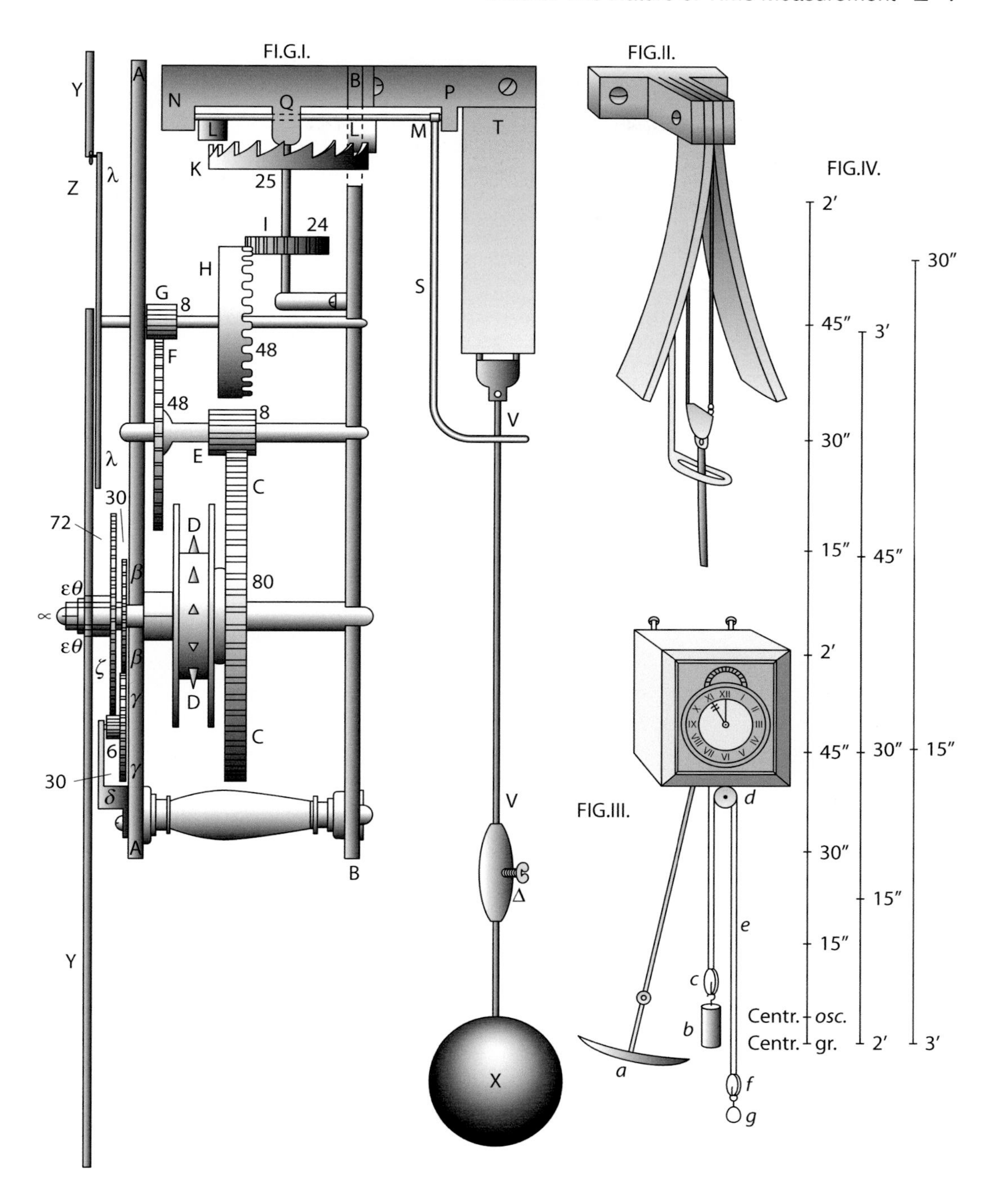

Figure 1.4 Diagram from the book Horologium Oscillatorium by Christiaan Huygens published in 1673 illustrating how one of his pendulum clocks was constructed. This clock used a form of escapement which differs slightly from the drop escapement illustrated in Figure 1.5, but the principle is the same. For more details on how the verge escapement illustrated here works, see https://en.wikipedia.org/wiki/Escapement Figure from https://ihttps://en.wikipedia.org/wiki/Horologium_Oscillatorium# /media/ File:H6_clock.jpg in the public domain.

Figure 1.5 Illustration of how the drop escapement in the pendulum clock works. For an animation see https://en.wikipedia.org/wiki/Escapement. The vertical lever is attached to the pendulum. The wheel is pulled in a clockwise direction by a cord wrapped around its axis and attached to a weight (or sometimes to a spring). As the pendulum swings to the left, the leftmost arm of the escapement slides off the tooth it's touching and the pendulum gets a push in the leftward direction from the wheel. Then the pendulum swings to the maximum to the left while the right arm of the escapement moves down and stops the motion of the wheel. Then, as the pendulum swings back right the wheel is released again until the left arm of the escapement comes down and catches it. Then the process repeats. The wheel is kept moving at a regular rate by the pendulum while the pendulum is kept moving against the forces of friction by the periodic pushes from the wheel. Used under the Creative Commons CC0 1.0 Universal Public Domain Dedication.

when the clock is wound (thus increasing the energy of the weight) and then the energy is fed back into the clock as the weight descends and the clock runs. The reason that a source of organized energy is necessary has its basis in the second law of thermodynamics (Chapter 3). In the case of the clock this is manifested in the fact that, without a weight pulling a shaft around, the clock would quickly stop running due to the action of friction in the mechanical parts of the clock, which rub against each other as they move. The friction results in the warming of the metal parts causing the clock to stop when there is no weight or when the weight has descended as far as it can. In energy terms, the energy of movement of the parts of the clock is converted into heat energy, which is not 'organized' in the sense intended here. Secondly, it is not sufficient, though it is necessary, for a mechanical clock to keep turning. The clock shaft must also turn at a steady rate. In the clock in Figure 1.4, this is assured by the pendulum. As noted, Galileo is credited with being the first to notice that a pendulum swings at a steady rate (as long as it doesn't stop due to the action of friction). This is actually a little bit surprising given what I have just said about the action of friction, but it is true that, if the friction is not too strong and the pendulum does not swing too widely, then the time between swings (the period) stays the same to a very good approximation, even though the distance (the amplitude) that it swings in one period gets smaller due to the action of friction (and in the absence of a weight pulling on the shaft).

So now we have a source of 'organized' energy (physicists say 'free energy') in the suspended weight and a means to keep a steady beat (the pendulum). The trick is to link them together so that the shaft keeps turning at a steady rate. This is achieved with what is called an 'escapement' of which there are many clever types. One of the simplest ones is illustrated in Figure 1.4. A string attached to the weight pulls all the time on a shaft which in turn causes a force on one of the toothed gears which is in turn transmitted to one of the two levers attached to the pendulum. Starting with the pendulum displaced to the right from the vertical, it swings left until the arm on the left transmitting the force to the pendulum slips off the toothed gear allowing the shaft on which the clocks hands are attached to move quickly forward under the action of the weight, while the pendulum continues to swing left, allowing a second arm attached to the pendulum on the right to engage the toothed wheel on the clock shaft, stopping the fall. The pendulum then swings right, releasing the wheel as the right arm disengages and the left arm again catches the wheel. Note that the pendulum gets 2 'pushes' per cycle from the arms.

We see in this analysis that mechanical clocks involve two empirical facts about the physical world which seem to some extent to be in contradiction: On one hand, there seem to be regular temporal processes in nature, such as the swinging of the pendulum and the movement of the planets. On the other hand, there are natural processes which result in things slowing down and stopping, such as the action of friction in the works of the clock. Our intuitions about time and certainly our ways of measuring time arise mainly from the regular processes. One sees that from the earliest human history known, in which time measurements were focused on understanding of the movements of objects in the sky. But the relation of these regular movements to the countervailing tendency of things to slow down and stop moving has been a recurrent theme in the effort to understand time. In medieval times in Europe the situation was rationalized by the idea that the 'heavenly' motions were 'perfect' while on earth no regular motions were possible and everything quickly stopped moving. Aristotle's physics, rediscovered in Europe from Greek times through Arabic libraries and the dominant view of earthly physics in Europe until the seventeenth century, took this view. Thus constructing an earthly clock, in which the regular motions were, in a sense, 'brought down to earth' had slightly sacrilegious implications.

Pendulum clocks nevertheless regulated religious life in monasteries and convents as well as business in commercial towns from the seventeenth century on. In the early modern era in Europe, long voyages of discovery led to an urgent need for better clocks for use in navigation. Determining latitude to a good approximation could be done by measuring the angular distance of the north star above the horizon. No accurate determination of time was needed. However to determine longitude, the only reliable method was to determine time as measured in a standard place, usually Greenwich, England when the sun in your position was at its highest point (local apparent noon). Then, using the same reasoning as used to describe Ricolli's measurements in Figure 1.3, one could determine the longitudinal angle relative to that in Greenwich, because the earth is turning (about 15° per hour). A famous royally sponsored contest for the best clock (chronometer) occurred in Great Britain in the 18th century starting around 1710 in an effort to meet navigation's need to accurately determine longitude. Huygens submitted his clock as a candidate for that prize. Throughout the history of time and timekeeping, the drive to measure time more accurately has often been driven by the needs of agriculture, shipping by sea, telegraphic communications, railroads, airlines, the global positioning system and other transportation and communication systems. For more discussion of the longitude problem see [43].

1.3 BIOLOGICAL CLOCKS

The clocks described so far have not depended on processes in living things. However, though most scientific work has depended on such nonliving clocks, there is strong evidence that almost all living organisms have an innate sense of time. This was noticed early on and some philosophers, such as Kant in the 18th century, suggested regarding the instinctual human sense of time as fundamental. Here we will review a little of what is known about biological clocks throughout the terrestrial biosphere. The discussion can be divided into what is known by observation of the behavior of whole organisms (including plants) and, secondly, what is beginning to be understood about the molecular basis for the functioning of these biological clocks.

Clocklike behavior in organisms occurs on time scales from milliseconds to years and the corresponding scientific study of it has divided into specialties such as the study of circadian, or daily, rhythms. The whole field has recently become known as chronobiology. The most familiar short-term biological periodicity is the beating of the heart, which is characteristic of vertebrate animals and insects though not of plants or microrganisms. At the macroscopic, whole organism level, experiments have been done with humans, for example, by confining them (voluntarily) to spaces in which they do not observe the cycles of light and darkness characteristic of the terrestrial day. They nevertheless maintain a schedule of eating, sleeping and other activities which conforms quite closely, but not exactly, to a 24 hour periodicity over experimental times of many weeks to months. On a longer time scale, experiments with migrating birds have shown that when confined to cages in which they get no external signals which would tell them that the seasons are changing, they nevertheless demonstrate activity which points strongly to their inclination to fly on their ancestral migration at the time characteristic of their specie's migrational pattern. Similar experiments show that hibernating mammals do not depend on external signals for timing the annual pattern of extended sleep which they exhibit.

The molecular mechanisms of some biological clocks are starting to be sorted out in great detail. For example, in the case of the circadian (daily) rhythm in the fruit fly Drosophila, a cycle of chemical changes in the protein chemistry of the cells of the organism takes about a day according to the model illustrated in Figure 1.6. (Appendix 1.2 reviews some basic information about proteins and transcription of the codes for proteins from the genetic material in the nucleus. However this is not required for gaining a qualitatively correct impression from this figure.) In the figure and caption, the entity PER is a protein. A portion of the DNA of the drosophila which codes for the protein PER is indicated by a wavy black line in the nucleus of a cell of the fly. In the early morning, messenger RNA molecules (wavy green line) begin to be copied from that DNA code. A little later the messenger RNA molecules leave the nucleus and find a ribosome in the rest of the cell, where the code in the RNA is used to make copies of the protein PER (yellow dots in figure). This first part of the sequence is universally the way the code on the DNA is translated into proteins. The next part is special to the clock mechanism: The PER proteins find their way back into the nucleus in the early evening and attach themselves to the DNA in a way which stops the production of the RNA containing their code. Thus the PER molecules halt their own production. (This is known as negative feedback.) However, the PER molecules are short lived and get disassembled by a process called phosphorylation so that most of them disappear rather quickly (indicated by the X in the figure) leaving the DNA to start producing the RNA again in a repeat of the cycle. See [48] for more details and references. (Jeffrey C. Hall, Michael Rosbash and Michael W. Young won the 2017 Nobel Prize in physiology for their studies of this circadian mechanism in Drosophila [48].) Very similar chemical circadian clock mechanisms are being found throughout the biosphere, for example, in fungi, rodents and humans, though the details differ from one species to another.

1.4 DEEP TIME AND ITS MEASUREMENT

All these clocks up to and including our modern atomic clocks measure short spans of time, up to years. Another line of development in time measurment concerns the distant past. When scientists state, for example, that the dinosaurs died out about 65 million years ago, that the earth is about 4.5 billion years old or that the universe is about 13.8 billion years old, what kind of measurements lead to these statements about 'deep time'? There are several methods. I will briefly discuss 1) paleontology and related methods 2) 'dating' using radioactive isotopes 3) extrapolations from the rate of expansion of the universe. These methods are actually used in the three exemplary cases of the time since the dinosaurs, the age of the earth and the age of the universe.

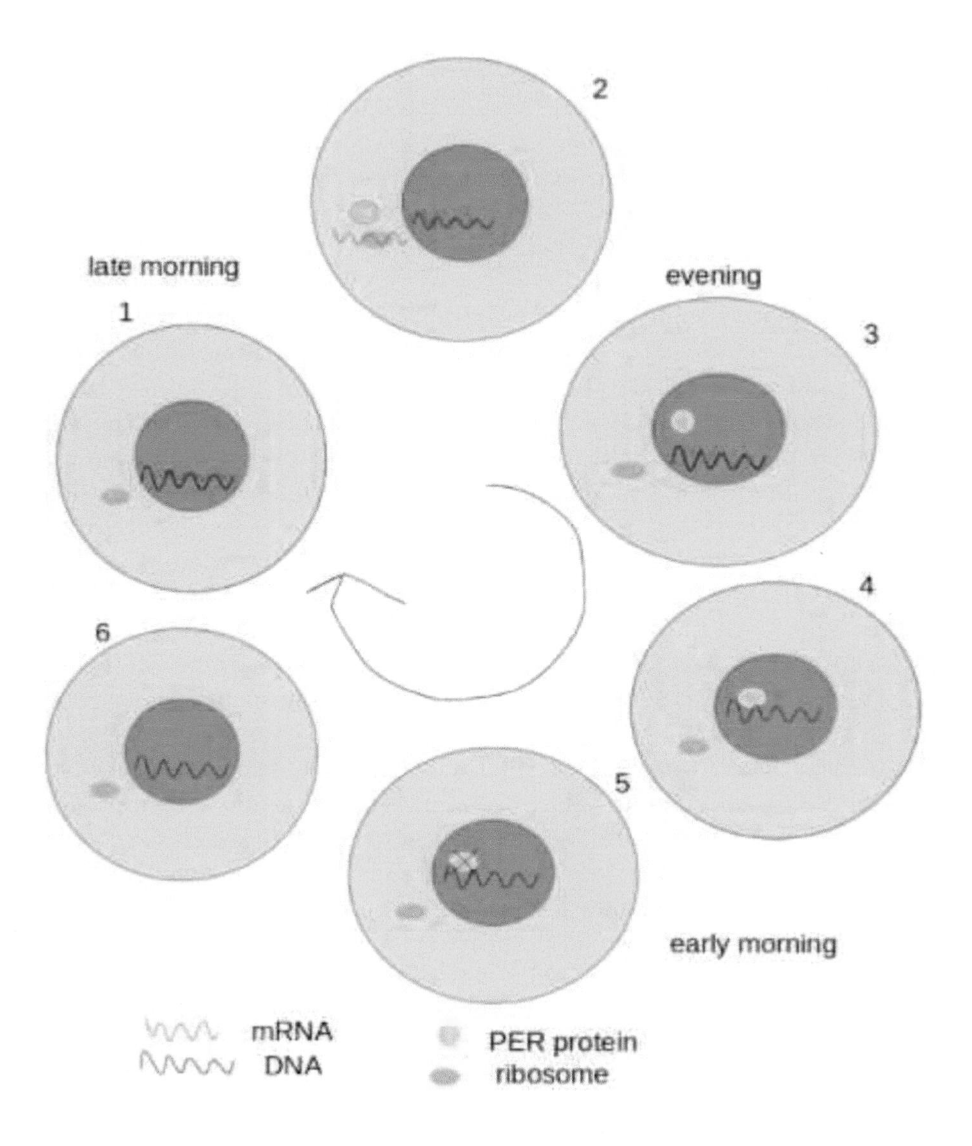

Figure 1.6 Model for the regulation of circadian transcription in Drosophila. At 1, when the sun is shining, a portion of the DNA in the nucleus is read to produce messenger RNA as indicated by the arrow and the wiggly line. The messenger RNA leaves the nucleus and finds a ribosome which reads the messenger RNA and produces the protein PER. as indicated in 2. The protein PER accumulates outside the nucleus and then enters it with the aid of another protein called TIM (not shown) as indicated in 3 in the early evening. The protein PER then attaches to the DNA and stops the production of the messenger RNA for its own production as shown in 4. In the early morning the PER is destroyed by phosphorylation, indicated in 5 and 6, and the production of messenger RNA for its production resumes in 1. For more details see [13] and [48]

1.5 PALEONTOLOGY AND PHYLOGENETIC TREES

Many Greek and Roman philosophers of the period from 3000 to 2000 years ago imagined the history of the earth to extend infinitely into the past, though some of them thought it was cyclic. However in Europe in the Middle Ages the earth was assumed to have had a very short life of the order of 10,000 years. The realization that the age was much greater occurred in the late 18th and early 19th centuries with the discovery of many fossils of large extinct mammals and reptiles all over the world but very significantly in North America. The discoveries of fossils of large extinct animals caused tremendous controversy and debate and required at least half a century to resolve [26]. That was partly because two conceptual changes had to be accepted, involving the possibility of extinction of species and the magnitude of the age of the earth. The field of paleontology, which studies fossils, determines their age partly by carefully recording the depth of the layers of soil under which a given fossil is buried when discovered. This method provides a reasonable guide if, for example, there is annual flooding, so that each layer can be identified as corresponding to a year, and if the layers have not been disturbed, for example, by volcanoes or earthquakes. Unfortunately all those conditions are rarely met, and various corrections have to be taken into account. For example, we now understand that earthquakes are caused by the slow motion of the continents which are drifting in response to upwelling of material from the mantle of the earth at ocean trenches. The continents move at about 1 cm/year. The result is that over millions of years, fossils originally laid down in the ocean floor have been carried into modern mountains. The discovery of fossils of marine animals in mountains was one of the most surprising events in the paleontology of the 18th and 19th centuries and was not fully understood until quite recently. The dating of fossils in our era uses the traditional methods of counting layers supplemented by dating with radioactive isotopes, discussed below. Dating also uses information gleaned from the genetic material of present day organisms, which themselves carry a record of the evolutionary history of the organism which can be used to estimate the time since their ancestors lived. Figure 1.7 shows a timeline of the history of the earth as determined by these combined methods.

To get information about the relationship of modern organisms to each other and to their evolutionary ancestors, information about the biochemistry of living organisms can be used. I will illustrate with an example involving a protein called a ribonuclease which is found in the second stomachs of ruminants (deer, goats, etc) which eat grasses which contain a lot of hard-to -digest cellulose. In the second stomach, these animals digest bacteria which ate the cellulose in the first stomach. They do this with the help of several proteins including the one called ribonuclease which breaks down the bacterial nucleic acids (DNA and RNA). (A review of some basic facts about molecular biology appears in Appendix 1.2.)

To figure out the ancestral history of the ruminants, molecular biologists compare the series of amino acids in their ribonuclease proteins. If the series are very similar, then the animals had a common ancestor in the recent past, whereas if they are very different, the common ancestor is farther back in evolutionary time. In this way, the biologists can construct an ancestral 'family tree' for the ruminants, as shown in the

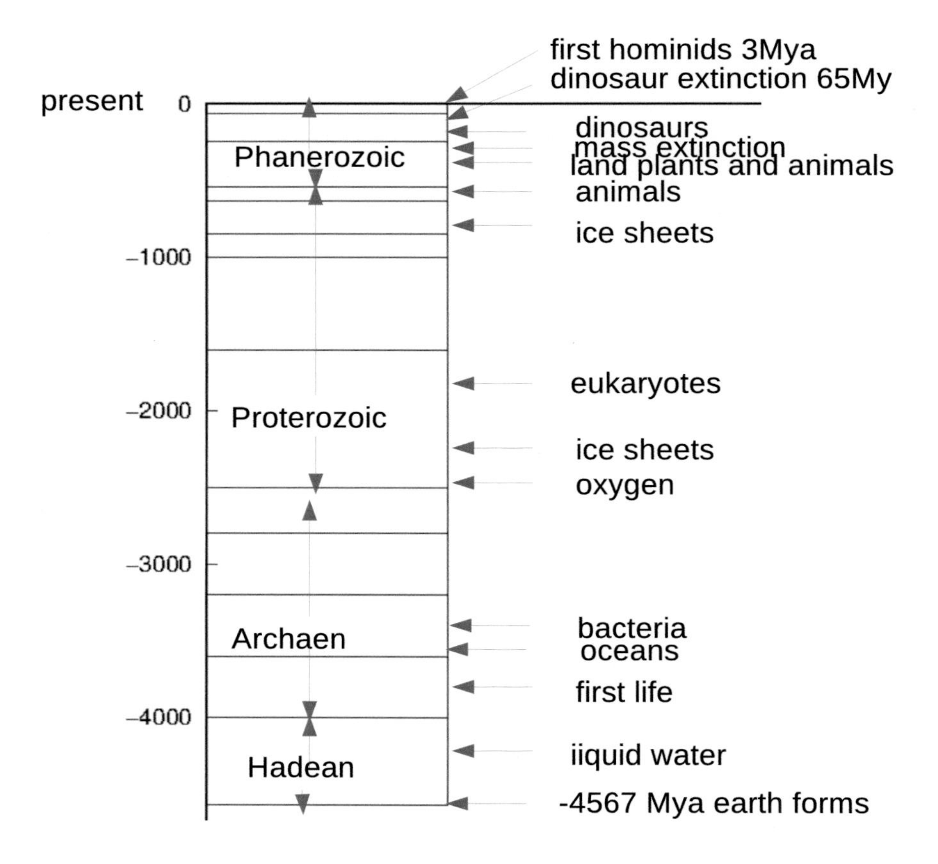

Figure 1.7 Ages of the earth as determined by paleontological, isotopic dating and biochemical methods. Mya means millions of years ago. The time that there have been humans on earth is less than a million years and is too short to show graphically here. Eukaryotes are organisms whose cells contain nuclei containing (most of) the cell's DNA. The first multicellular organisms appeared about 300 million years ago. The gas containing the diatomic oxygen molecules which humans breath and use for metabolism did not appear on earth until more than 2 billion years of its history had passed.

Figure 1.8. This is how they got the form of the tree, but how did they get the time scale on the bottom? That was by using a combination of paleontological data and isotopic dating of layers (described below) to establish when ruminants started to appear in the fossil record. It turns out that this happened about 35 million years ago in the paleontological epoch called the Oligocene, when other kinds of isotopic data establish that earth's surface began to cool. (Just before that the earth was

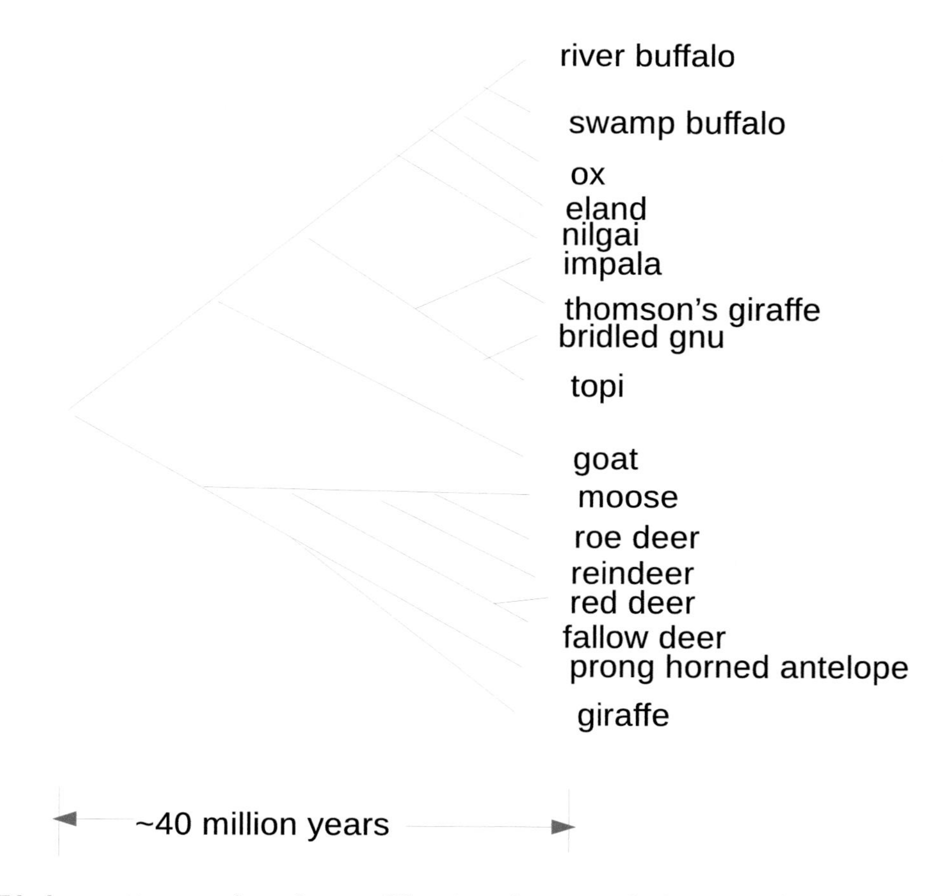

Figure 1.8 Phylogenetic tree of ruminants. The time shown at the bottom of the picture was estimated using radioactive dating. Redrawn from [19].

largely a tropical rain forest.) Cooling led to a lot of grassy plains, instead of rain forest. The ruminants could eat the grass in those plains, and their predecessors could not.

The older way to decide how closely species are related is to compare their traits, called their phenotype, and trees were constructed for a long time, starting with Linneaus in the eighteenth century, using that method. Often, the two methods are used together, particularly because dating is easier with fossils and it is easier to find all the stages with genotype studies. (Fossil records have 'missing links'.) In Figure 1.9 some trees derived with the two methods are compared.

1.6 DATING WITH RADIOACTIVE ISOTOPES

All the material on and near the surface of the earth (and much of the cold matter elsewhere) consists of combinations of atoms of the chemical periodic table. These atoms consist of a nucleus which is very small compared to the size of the atom as a whole and which, oddly, contains most of the mass of the atom. The nucleus carries a positive electrical charge arising from the protons within it. The atomic number Z of the atom is the number of protons in the nucleus. In addition the nucleus contains neutrons, which have about the same mass as the protons but carry no electrical charge. There are A-Z neutrons in the nucleus where A is called the atomic mass number. Around the nucleus, when the atom is electrically neutral, Z negatively charged electrons orbit. Each electron has an electrical charge which is exactly the same size as the charge of a proton, but opposite in sign, so that the charge of the neutral atom comes out zero. The electrons have very little mass compared to the nucleus but the size of their orbits determines the size of the atom, which is about 100,000 times bigger than the size of the nucleus. Everyday chemistry as it occurs in living systems, cars, cooking etc only involves rearrangement of these electrons as the atoms interact with each other. Atoms with the same Z (number of electrons), therefore, all behave chemically in the same way and are considered the same chemical element. But atoms of the same Z can have different numbers neutrons which don't affect the ordinary chemistry but do affect the mass. Atoms with the same Z but different A are said to be different isotopes of the same chemical element. We encountered an example in the description of the cesium atomic clock: In the designation "cesium 133 atom" 133 is the value of the mass number A. The atomic number Z of cesium is 55. The cesium (Cs) 133 nucleus never decays: we say it is stable. One other fact is relevant. Some nuclei are not stable. Over times which are characteristic of each isotope they decay into other nuclei. Currently known information about all the known isotopes is collected in online tables such as [34]. There are many consequences of this instability of nuclei including nuclear generation of electrical energy and nuclear weapons but for our purposes here it is relevant that this feature of nuclei makes them usable as clocks for measuring times in the distant past.

As an example, which is historically one of the first cases of this kind of use of isotopes, we consider the isotope ^{14}C of carbon. Carbon is the chemical element with atomic number Z $=$ 6 and the 14 in the notation is the value of A so the nucleus ^{14}C has $14 - 6 = 8$ neutrons. It turns out that ^{14}C is not stable and decays into ^{14}N plus an electron and another elementary particle called an anti-neutrino. The decay occurs in about 5730 years. What I mean by 'about' is that, if you have a gram of ^{14}C atoms, then in 5730 years, half of it will be gone and if you wait another 5730 years half of that half which was left will be gone and so forth. The number 5730 years is called the 'half life' of the isotope. Now to understand how these facts are used to make a clock, you have to know that in the atmosphere of the earth, which is mostly nitrogen, ^{14}N nuclei are constantly being bombarded with high energy neutrons (from cosmic rays from outer space) resulting in a ^{14}C and a proton. Thus ^{14}C is being continuously created in the atmosphere. The creation rate balances the decay rate on average so there is continuously a fixed amount of ^{14}C in the atmosphere. When plants grow, they absorb carbon in the form of carbon dioxide and thus growing plants also contain carbon containing the same fraction of ^{14}C which is found in the atmosphere.

After the plants grow, they die and their remains are buried beneath a variety of debris such as other plants, silt from floods, lava from volcanoes and so forth. Over centuries, the overburden is sufficient to shield the remains from the cosmic rays, so the creation of ^{14}C stops. But the decay of ^{14}C continues. So 5730 years after burial, 1/2 of the original ^{14}C is left and so forth. Therefore, by measuring the amount of ^{14}C in buried plant material and comparing it with the amount in the atmosphere, one can get a rather good estimate of how long the material was buried. The method was developed in the mid twentieth century and confirmed to be accurate by dating archaeological artifacts including a royal Egyptian barge which was buried in 1850 BCE. It has since been used particularly for the dating of artifacts in the ruins of ancient human settlements. Much of the early work on dating using ^{14}C was done by Willard Libby and coworkers [37]. Libby won a Nobel prize for his efforts in 1960.

Another isotope in common use is ^{238}U (uranium, atomic number 92). It decays by emission of a ^{4}He nucleus (called an alpha particle for historical reasons) with an extremely long half life of about 4.46 billion years. A rather long series of decays occurs after that first one, but they are all much faster and the nucleus ends up as ^{206}Pb. (Pb is lead, atomic number 82.) Effectively, the half life of the ^{238}U decaying to ^{206}Pb is about 4.46 billion years and this makes it useful for determining the time of events very far in the past. By comparing the amount of the lead isotope to the amount of ^{238}U in deposits of uranium on the earth, one can estimate the amount of time which has passed since the deposit was all uranium. It was in part by use of this method, with some refinements as described in Appendix 1.3, that the age of the earth has been determined to be about 4.5 billion years [45].

A third isotope in common use for dating events in the deep past is ^{40}K. (K is potassium, $Z = 19$.) Its half life is 1.25 billion years, and one can date the age of volcanic deposits using it by monitoring the amount of the decay product, ^{40}Ar (Ar is argon, $Z = 18$) because the argon escapes from molten rock but gets trapped after the lava freezes [12]. Thus, by measuring the amount of ^{40}Ar trapped in the frozen lava, one can estimate the time since the volcano occurred and the lava froze.

One sees from this brief review that the use of radioactively decaying isotopes as clocks for measuring events at times far in the past has some elements in common with the use of cesium atoms for establishment of a short time unit. In particular it is implicitly assumed that all the nuclei of a given isotope are identical, similar to the assumption that all cesium atoms behave in the same way. Again, there is strong evidence for the consistency of this assumption with our experience but absolute proof is not possible. We can again ask whether these isotopic clocks are determining the times since ancient events occurred or are, on the other hand, defining what we mean by the times since the events occurred. However the times determined by isotopic dating have been shown in many cases to be consistent with times determined by other methods associated with astronomical periodicities such as the period of the orbit of the earth around the sun and with the fossil record. The consistency strongly suggests that something is being measured which is not only associated with the properties of particular isotopes.

1.7 USING THE UNIVERSE AS A CLOCK: THE AGE OF THE UNIVERSE

For even longer times, scientists use features of the observed universe as a whole to estimate deep times. In some respects this is an astronomical method because one is using observations of phenomena in the sky. But the details are very different from those characterizing methods based on observations of the planets of the solar system which were used in the 18th through 20th centuries and earlier to define units of time. The stars observed in the sky can be roughly divided into two groups: Those nearby lie in a disk visible on dark nights without urban light pollution as the 'Milky Way'. Those stars, roughly 10^{11} in number, are said to be within our own galaxy. However most of the stars in the sky are in other clusters also called galaxies, some like our own and others quite different. Galaxies were first discovered in the early twentieth century, with significant contributions by Edwin Hubble. To understand how information about light from galaxies can lead to estimates of the age of the universe, the key observation is that, beginning with the work of Hubble in the 1920's, the colors of the light from the stars of distant galaxies were observed to be reddened in a way that allowed a determination that the galaxies were (and are) receding from us. The determination is based on the Doppler effect (reviewed in Appendix 1.4) together with the assumption that the atoms of the elements in the galaxy emit light at characteristic wavelengths which are the same as those of the corresponding atoms on earth. What Hubble and his successors in this work found is that the velocity with which a galaxy is receding from us is larger, the farther it is away from earth. The relation was found to be linear. That is, $v = Hr$ where r is the distance between us and the galaxy and v is the speed with which it is receding. It should be clear that the constant H (named for Hubble) has the units of 1 over time.

To see the significance of it, we refer to Figure 1.9: Consider 2 observed galaxies at the present time, at present distances r_1 and r_2 from us. If the relation $v = Hr$ is true for each galaxy at the present time, then we can show that each galaxy was at zero distance from us at a time which is $1/H$ earlier than the present. The relation is illustrated in Figure 1.10. This is the basic notion which has led to the accepted scientific view that all matter in the universe was extremely dense about $1/H$ ago. (This does not mean that the universe was compressed to a point. In the current models, the universe is infinite, though we can only observe a finite part of it, the 'observable universe', because of the finite time required for the light from distant matter to reach us. At the time of the Big Bang the observable universe is thought to have still been finite but the density of matter in it was extremely large.) Some further details of how measurements of stars called supernovae have been used to refine this picture appear in Appendix 1.5 With the assumptions used here, the time turns out to be about 16 billion years. We may say that this determination is equivalent to using observations on the universe as a whole as a clock.

We have assumed here that the recession speeds have maintained the same relation since the early stage of confinement. That assumption is almost certainly not correct and has been corrected to give a more accurate estimate of the time since the big bang. An important correction arises because it is well-established that all

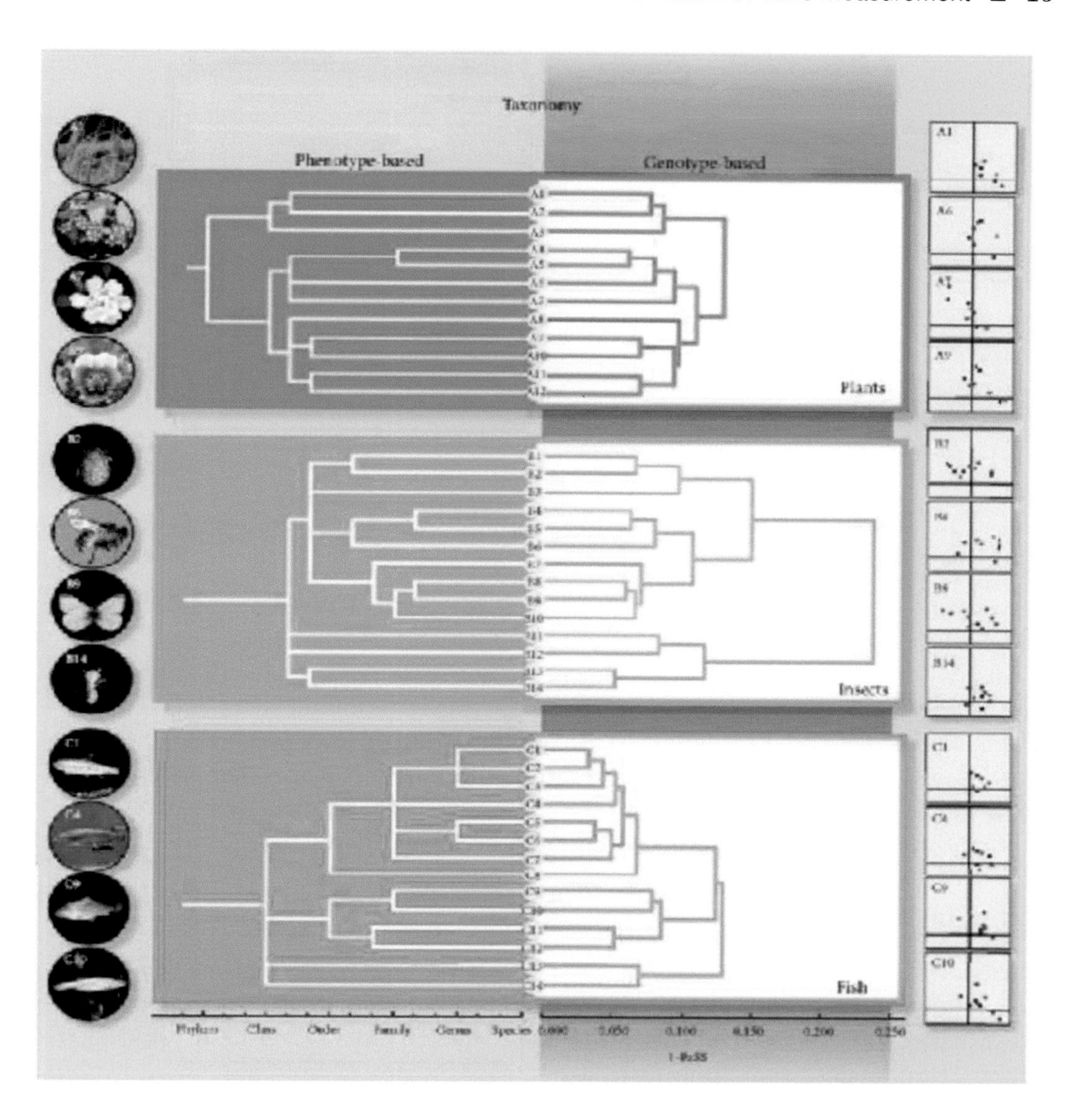

Figure 1.9 Phylodendrons of plants (A1-A12), insects (B1-B14), and fish (C1-C14). Phenotypic (left) and genotypic (right) trees are drawn on the basis of analysis of organism external traits (phenotype) and molecular biological features (genotype), respectively. From reference [18].

massive matter attracts other massive matter and, as a consequence, the exploding material should be pulled back toward the original tightly confined density and correspondingly decelerated. The effect is shown in Figure 1.11, in which the average distance between galaxies is plotted into the past. The various curves show how the straight lines of Figure 1.9 are modified if one takes the mutual gravitation of matter and the resulting deceleration into account, with various assumptions regarding the total amount of mass in the universe. A sufficiently massive universe would be

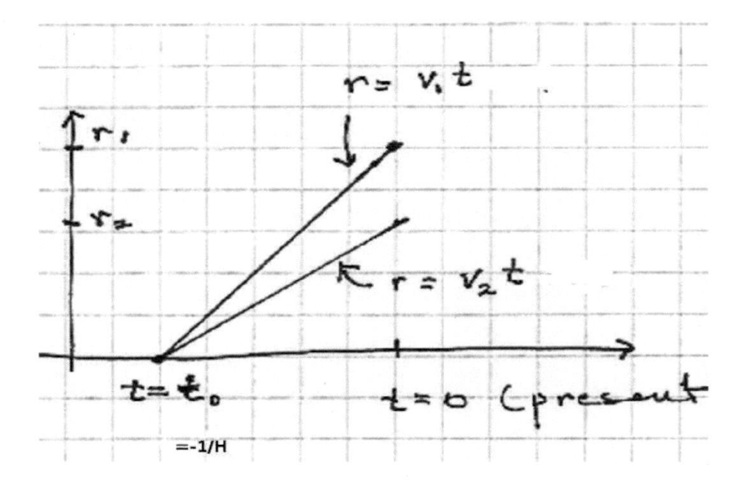

Figure 1.10 Illustration of the relation of the Hubble constant to the age of the universe.

predicted to eventually collapse back in what is sometimes called a 'big crunch' but the evidence does not support that possibility. Using current understanding of the evidence, the estimate of the time since the big bang is somewhat less than $1/H$ if one takes the deceleration into account: $1/H$ is about 16 billion years and the current estimates are around 13.8 billion years. Some details of this picture are still being worked out and some of the evidence used to obtain the estimate comes from observations of the background radiation observed at microwave frequencies by astronomical observations from satellites which I am not describing here.

1.8 SUMMARY

Here we have approached the question of the meaning and nature of time by describing how humans measure it. This provides a kind of implicit definition of what humans mean when they think and make statements about time. The technological history makes it quite clear that from earliest history, the idea of time has arisen from the observation of regular, repeatable processes observed by humans in their environment. For most of history, the most striking regularities were in the sky, and the motions of objects in the sky provided the standard for time measurement from early days among civilizations in all parts of the world to the mid twentieth century. Mechanical regularities such as the periods of pendula, were more convenient for use in clocks, but the fundamental calibration was always made from astronomical observations. More recently, regularities in the submicroscopic atomic regime have proven more regular than astronomical motions and provide the scientific standard in atomic clocks.

In another way, humans and other organisms of the terrestrial biosphere 'measure' time through their intuitive sense of it, arising from biochemical processes in the cells of their bodies which are just beginning to be understood. This biochemically based intuition of time is related to another line of thinking about time, particularly important in philosophy of the late 18th and 19th centuries, that regarded time as a fundamental intuition which could not be further defined.

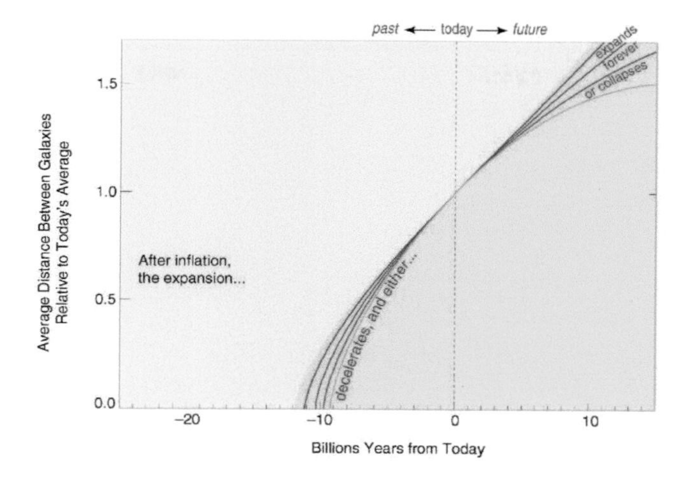

Figure 1.11 'Scale factor' versus past times in various theoretical models of the universe which take account of the mutual gravitational attraction of all matter. If the universe did not contain any matter these same models would give a straight line like the ones in the previous figure. The 'scale factor' is the ratio of the average distance between galaxies divided by the value of that average at the present time (t = 0). This figure is a part of the Nobel Prize lecture of Saul Perlmutter. Copyright Ī The Nobel Foundation 2011.

To measure 'deep time', back to the age of the universe, regularities in natural phenomena, particularly the half-lives of unstable nuclei provide the needed regularity to permit measurement. Other information concerning 'deep time' is gleaned from the depth of buried fossils as found by paleontologists, in comparison of genomes between living species and in the observed recession speeds of distant galaxies.

A feature of the use of natural regularities to define standards of time measurement is that not all natural processes are regular in the required sense. In fact it requires substantial effort to find such regularities and many processes exhibit temporal irregularities which makes them unsuitable as clocks or time standards. For example, we saw the interplay of regular motion and the dissipation of the energy of that motion through the action of friction in the design of mechanical clocks. The relationship of the regular processes to the countervailing tendency of things to slow down and stop is a recurrent theme in the efforts to understand time as we will see in discussions of time reversibility and thermodynamics in the next chapters.

CHAPTER 2

Issues in the Nature of Time

What then is time? If no one asks me, I know: if I wish to explain it to one that asketh, I know not.. Saint Augustine, 397 CE, Chapter XI translated by Edward Bouverie Pusey Oxford: J.H. Parker; J.G. and F. Rivington (1838)

2.1 INTRODUCTION

If time is what we measure with a cesium atomic clock, then perhaps no further discussion is required. However, even though some analogous statement has been possible for many centuries (eg "time is measured by the phases of the moon"), many questions have been raised about its nature. Is time a substance? Is it a label we put on events? If so, is there time when nothing happens? Is there time even if there is not material stuff present? Somewhat similarly, was there time before the Big Bang which seems to be the first observable event in our universe? Does time have a beginning? An end? Is it discrete or infinitely divisible? Why does time seem like a busy one way street? That is, why can time apparently not be stopped, reversed or accelerated? Or can it? Are there other definitions of time besides the one using a cesium atomic clock which mean something different? Why are we apparently always in the present with only memories of the past and hopes and fears for the future? Are the past and the future real? Maybe the past is real and the future isn't yet? What is the present anyway? Is it an infinitely thin slice of time and, if so, how can anything happen in it?

It turns out that to some extent the discussion of such questions depends on the perspective of the questioner, and that not every perspective is associated with exactly the same definition of time. For example, the questions refer repeatedly to 'events' though a little thought reveals that it is not easy to formulate a concise and general definition of exactly what an 'event' is. In practice, a working definition along the lines of "A phenomenon occurring in an infinitesimally short interval of time whose occurrence could be inferred from humanly detectable signals caused by electromagnetic or gravitational forces is termed an 'event'" will be assumed here. However, the dangers of both circularity and anthropocentricity are evident and I will not discuss them very much further here while nevertheless assuming that, finally, the history of everything is a history of 'events'. (Such definitions often also imply an

DOI: 10.1201/9781003037125-2

assumed model which contains postulated answers to some of the questions listed.) The measurements we have discussed in the last chapter can be described as measuring "physicists' Newtonian time". Different perspectives will arise if we consider "human intuitive time" or "biological time" (as experienced by many living organisms not limited to humans). Still further points of view arise from mathematics, the physics of heat, (technical) philosophy and relativistic physics. We will discuss each of these perspectives separately, but returning in each case to what it adds to understanding of questions such as the ones listed in the preceding paragraph and others related to them.

2.2 PRENEWTONIAN PERCEPTIONS OF TIME

Philosophers sometimes refer to everyday perceptions of time by nonphilosophers as 'folk philosophy'. An important point made by Whitrow is that the assumptions made by most humans about time in 2019, at least in highly technical societies, are quite different from those adopted by humans in the Middle Ages in Europe (1000 to 500 years ago), in the Roman Empire (2000 years ago) or among the Greeks when many philosophers from that region wrote and taught about time 3000 to 2500 years ago. For example, before the 'Christian era' time was assumed by many writers, and presumably by a wider public though that is less well known, to be cyclic. That is, time was thought not to have a beginning or an end, but events in time were thought to repeat themselves. Many writers thought that this repetition would be EXACT. Whitrow [60] quotes Nemesius from the fourth century AD who wrote that the Stoics of that and earlier times believed that the cosmos would be "restored anew in a precisely similar arrangement as before... each individual man will live again, with the same friends..." and so on in great detail. Notice that if the restoration were really exact in that way, it would be impossible for any observer in it to distinguish one reincarnation of the universe from the next. It would be difficult (perhaps not impossible), even today, for modern humans to prove that such a recycling is not occurring on very long time scales.

A significant shift in perspective on the nature of time occurred in Europe around the beginning of the current era as the Roman Empire was becoming Christian. The confusion which ensued is represented by Book XI of the Confessions of St Augustine, written by Augustine of Hippo, who lived from 354 AD to 430 AD in what is now Algeria, then a province of the Roman Empire. Augustine's mother was a Christian but he did not convert to Christianity until 386 when he was 32. Before that he had attended several schools and had been appointed to a prestigious position as professor of rhetoric in Milan, Italy. When he converted, he abandoned that career and became a priest, and was posthumously declared a saint, in the Christian church. But book XI of his famous 'Confessions', though containing frequent references to religious entities, is not mainly about religion but is a philosophical and scientific inquiry into the nature of time. In it, he describes the issues which arose concerning the nature of time in that epoch and how the proposed answers were changing in the transition from the preChristian to the Christian era. Augustine addressed the following issues: Does time have a beginning? What is the nature of the past, present and future? Are the past and future 'real' and, if so, in what sense? How should and do humans measure time and what are they measuring? With regard to the first question, Augustine concluded on the basis of his new religion that time did have a beginning, when God created everything including time. That was in quite sharp contradiction to earlier ideas of cyclic time or time which extended indefinitely far into the past and the future.

In discussing the present, Augustine wrote (in an old translation [51] from the original Latin): "See first, whether an hundred years can be present. For if the first of these years be now current, it is present, but the other ninety and nine are to come, and therefore are not yet, but if the second year be current, one is now past, another present, the rest to come. And so if we assume any middle year of this hundred to be present, all before it, are past; all after it, to come; wherefore an hundred years cannot be present. But see at least whether that one which is now current, itself is present; for if the current month be its first, the rest are to come; if the second, the first is already past, and the rest are not yet. Therefore, neither is the year now current present; and if not present as a whole, then is not the year present. For twelve months are a year; of which whatever by the current month is present; the rest past, or to come. Although neither is that current month present; but one day only; the rest being to come, if it be the first; past, if the last; if any of the middle, then amid past and to come. See how the present time, which alone we found could be called long, is abridged to the length scarce of one day. But let us examine that also; because neither is one day present as a whole. For it is made up of four and twenty hours of night and day: of which, the first hath the rest to come; the last hath them past; and any of the middle hath those before it past, those behind it to come. .. [no] instant of time [can] be conceived, which cannot be divided into the smallest particles of moments, that alone is it, which may be called present. Which yet flies with such speed from future to past, as not to be lengthened out with the least stay. For if it be, it is divided into past and future. The present hath no space".

Here Augustine is suggesting that the present must logically be an infinitely short interval of time, yet, as he discussed later, it seems that the present alone is real. A form of that paradox was posed by Zeno a millennium earlier [59]. The sense in which the present can be regarded as infinitely short was only clarified more than a millennium later when Newton and Leibniz invented calculus.

Turning to the question of the past, Augustine wrote:

"if times past and to come be, I would know where they be. Which yet if I cannot, yet I know, wherever they be, they are not there as future, or past, but present. For if there also they be future, they are not yet there; if there also they be past, they are no longer there. Wheresoever then is whatsoever is, it is only as present. Although when past facts are related, they are drawn out of the memory, not the things themselves which are past, but words which, conceived by the images of the things, they, in passing, have through the senses left as traces in the mind".

and of the future:

"we generally think before on our future actions, and that forethinking is present, but the action whereof we forethink is not yet, because it is to come. Which, when we have set upon, and have begun to do what we were forethinking, then shall that action be; because then it is no longer future, but present".

and finally

"perchance it might be properly said, 'there be three times; a present of things past, a present of things present, and a present of things future". For these three do exist in some sort, in the soul, but otherwhere do I not see them; present of things past, memory; present of things present, sight; present of things future, expectation".

This makes Augustine sound like a 'presentist', a school of contemporary philosophy which holds that only the present is real [58]. (In academic philosophy it is opposed to a contrary view termed 'eternalist'.)

But he recognizes that this implies a very puzzling contradiction, because "The present hath no space". yet is the only reality and if we want to measure time we have to do it in the present: "we measure times as they pass, in order to be able to say, this time is twice so much as that one; or, this is just so much as that; and so of any other parts of time, which be measurable. Wherefore, as I said, we measure times as they pass". But that too is puzzling: "But time present how do we measure, seeing it hath no space?" However, though Augustine's detailed descriptions of time measurement are technically very primitive, he does suggest a resolution to the apparent paradox just cited: To measure the time interval between two events, one starts some kind of clock (a device with a naturally repeating phenomenon such as a pendulum) and counts the number of repetitions of the clock cycle until the second event. All the actions are taken in the present and a semipermanent record of the result (the number of cycles) can be recorded, also in the present, either in memory or another medium (for example, paper).

I would not suggest so much that these ancient puzzles have been resolved by modern science as that they have been reframed. In physics, calculus makes it possible to describe an infinitesimally short present precisely, but the intuitive significance of the description may not be easy to grasp. That is further discussed in the next section (2.3) on Newtonian time. Regarding whether time has a beginning, in the middle of the last century physicists considered two models of the history of the universe: one, a steady state universe in which the mass loss associated with the receding galaxies was compensated by spontaneous appearance of more hydrogen and the other, now dominant, Big Bang model. The latter may be said to imply a kind of beginning to time. Contemporary time measurement as described in Chapter 1 follows procedures which can be seen to follow the general path described by Augustine. I review how science in the last two millennia has affected modern answers to the questions posed by Augustine in Chapter 6.

2.3 NEWTONIAN TIME

Newton lived more than a millennium after Augustine. His model of the world has proved immensely useful in describing astronomical events as well as engineering devices and much else on earth and is still used throughout the sciences. It can be said that he provided definite answers to several of Augustine's questions which, at least for a wide range of phenomena, permitted prediction of future events from data describing the present. Newton described time as a smoothly varying quantity, measured by a real number (not an integer) which increases continuously and cannot be stopped by any physical process. He referred (at least in some translations of his work) to its 'flow', suggesting a picture in which the observer is stationary and time flows past him. However the mathematics of his theories do not make much use of this idea of 'flow', which is difficult to make precise (flow with respect to what?), and could equally well be regarded as supporting an intuitive view in which the observer moves forward through time.

In Newton's model, the present is a single point somewhere on the continuous set of monotonically increasing numbers labeling times. Time values larger than the present value are future times. Time values smaller than the present value are past times. There is no reference to a beginning of time. The stuff of the universe in the Newtonian universe consists entirely of particles which are confined to sharply defined points in three-dimensional space. To approximate the modern world as we now know it with a Newtonian picture, the particles are most naturally identified with atoms. The number of such atoms in the observable universe is immense. A very rough estimate gives 10^{80}. In the Newtonian picture, one can imagine laying out the 'trajectory' of each particle in the universe as a function of time in a graph. People do things like that all the time with computers using Newtonian mathematics to describe gases, clusters of stars, biopolymers, weather and climate and much else, taking care to only use the Newtonian picture where it is now known to work well. (We will discuss limitations of the Newtonian picture in Chapters 4 and 5.)

Here is a simple example. I have shown the positions of three objects which you may think of as atoms on a flat surface, as shown in the Figure 2.1. A rule for calculating the forces between them for each set of positions of the three objects has been chosen. The atoms are placed in a not quite equilateral triangle on the surface with no velocities and then they are let go. It is quite easy, using a computer, to solve the Newtonian equations of motion for what happens to them after that. (Some details about how the calculation was done is given in Appendix 2.1.) In Figure 2.1, each particle's position is defined by the distances (x, y) of the particle from the origin in the picture. Figure 2.2 shows what happens to the value of y according to Newton for the set of starting positions shown in Figure 2.1.

When you look at that graph, there is little which distinguishes the past from the future except that one is to the left of the time point called the present and the other is to the right. It seems quite reasonable to regard all the states on the graph as real. The example thus provides an indication of why most physicists regard past, present and future events as equally real [11].

Another characteristic feature illustrated by the example is that the trajectories are definitely predicted, in principle (up to small errors arising from the limitations of the computer) exactly. If you believe that the universe is completely described by Newtonian physics, then that suggests conclusions about the idea of free will which are disturbing to many people and agitated both philosophers and others with inconclusive results from the time of Newton to the present. The problem is that there is a lot of evidence that the processes taking place in human brains and other body parts as well as in those of other animals are described by deterministic Newtonian

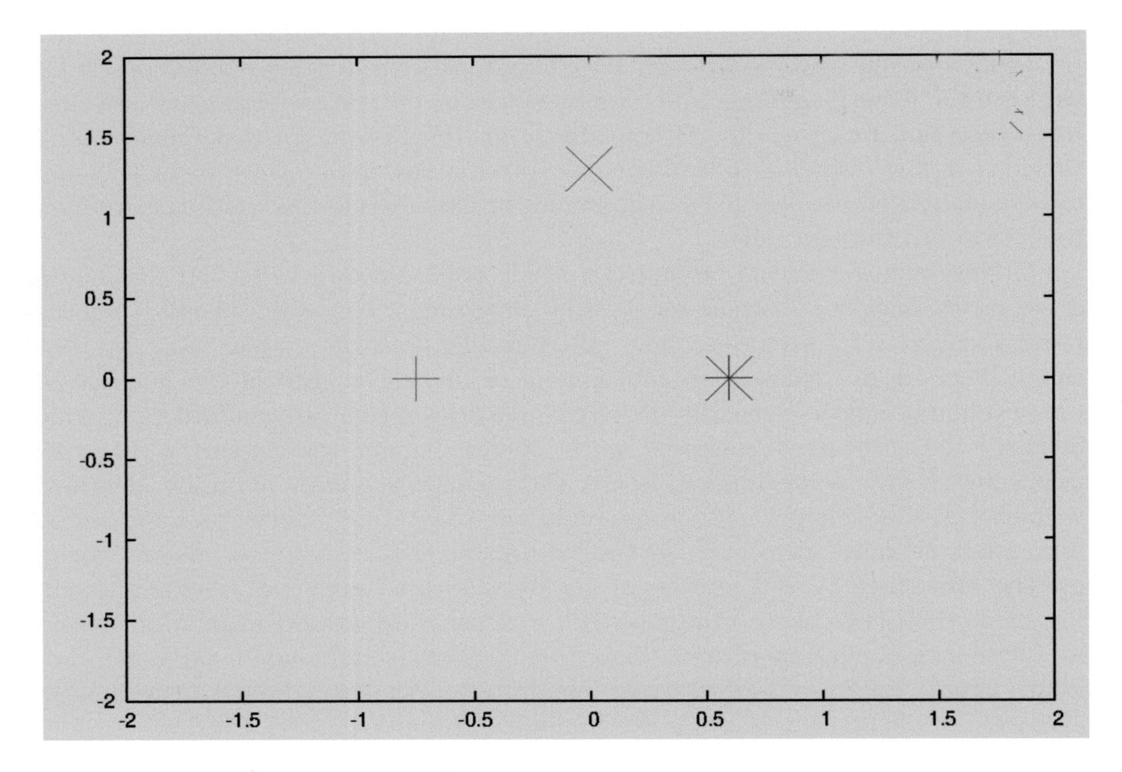

Figure 2.1 Starting positions of three particles on a flat surface.

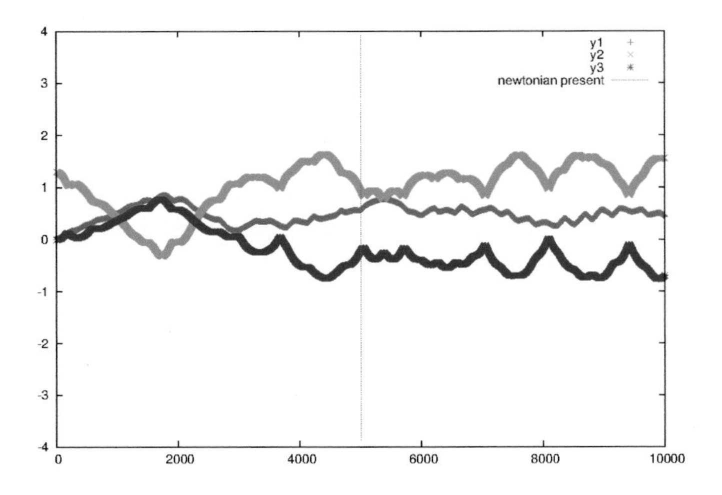

Figure 2.2 Newtonian trajectories of the vertical position of the three particles. All the positions are definitely predicted.

physics. That seems to imply that the choices humans make are also determined by their past and present states and the idea they make choices is an 'illusion'.

I will not attempt to reproduce several centuries of philosophical argument about the problem of free will here, but suggest the following perspective: Suppose we regard our brains as computational devices which have evolved to enhance our chances of surviving and reproducing. One way they do that is to take in information about the local environment and use it to plot alternative scenarios for future action, weighing the possible consequences of each scenario in light of available information including both information about the present local environment and memories of the past. The scenario judged to be the one most favorable is then chosen by the brain and some action is taken, eg by sending messages to legs to flee or arms to fight. A choice has been made, it was significant, but the process of absorbing information, processing and weighing it and choosing the most favorably weighted alternative could all have been deterministic, determined by the input information, the memories in the brain and its processing capabilities. This perspective is deterministic: the future is completely determined by the present and the past, but it does not suggest that the person or animal making the choice was acting like some kind of a robot, as sometimes is suggested. The individual (sometimes called the 'agent') was actively involved in the formulation of the possible courses of action and the determination of the choice made, through the neurological processes of conceiving, evaluating and weighing alternatives. Some details of how this process occurs are becoming available through the use of probes which record the electrical potentials in neurons of human brains while individuals are making simple choices [40].

A similar description of the process by which individuals are held socially responsible for their actions is also possible. The social rules of action are products of the evolution of human and other animal groups which permit those groups to survive and thrive. It has turned out to be expedient for that purpose to inculcate behaviorial standards in members of the group and then to punish or otherwise deter those that do not follow the rules. All the processes described could be (and probably are) deterministic in the Newtonian sense but the act of holding an individual responsible for following the rules would still be appropriate and understandable as a mechanism for enhancing the likelihood of group survival. For more detail concerning the philosophical debate about free will, which continues, see [53]. I will return to this in Chapter 3, where I suggest the perspective that consciousness and intent in living organism can be usefully regarded as emergent, macroscopic, coarse-grained phenomena with properties, such as apparent free intent, which are qualitatively only apparent at the coarse-grained level.

Another feature of the example of three objects illustrated in the figure, is that the paths they follow according to the Newtonian equations are 'time reversible'. What that means is that if, at some moment, one suddenly and exactly reverses the directions of all three velocities, while retaining their magnitudes, the three objects will, according to the Newtonian equations, exactly retrace their steps to the starting points. That can be illustrated by use of a code written using the methods described in Appendix 2.1. I will discuss this feature in more detail in Chapter 3, but note here that it is quite different from our daily experience, in which many events seem entirely irreversible. In Chapter 3, I will try to convince the reader that the two views can be reconciled.

2.4 THE NEWTONIAN AND THE HUMAN PRESENT

I will now consider the contrast between the Newtonian view and the 'folk' view of time in some more detail. Some, but not all, of the contrasts would be rendered less stark if I compared human experience with the picture presented by modern physics including significant modifications of Newtonian physics as a description of the world when masses are large, velocities are high or length scales are very small. However the comparison with Newton's world illustrates some general features which I think lie behind much of the confused debate about the nature of time.

When humans speak of and think about 'the present' in their everyday lives they are not thinking of a point of negligibly small time duration. Our mental processes are not instantaneous so that would be impossible. Measurements of response times of the human brain suggest that the fastest processes proceed in times of a few tens of microseconds (about a hundred thousandth of a second) at least. But in fact, in everyday talk, "the present" might mean "this hour" "this day" , "this year" or even "this decade or century". Further, when we speak of such a "present" we usually mean "up to now" , that is "this day so far" though that is not always the case: we might be including some prediction of the immediate future in what we mean by "this day". (Notice how similar that is to Augustine's perception of the present as quoted earlier.) In any case, the ordinary human "present" is a little piece of the Newtonian past up to, and perhaps a little beyond, the Newtonian present.

Looking a little more closely however, the difference between the Newtonian present and the intuitive human present is somewhat reduced: The state of the world in the Newtonian present, while regarded as representing the world at an infinitesimally small slice of time, cannot actually be determined without consideration of the immediate past. That is because the specification of the state requires the specification of *instantaneous* velocities. These days, most people think they have an intuitive understanding of instantaneous velocities, largely because of ubiquitous speedometers (which read the magnitude of instantaneous velocity) in automobiles. However, instantaneous velocity is a very subtle concept and it required many centuries for humans to get a clear picture of what it means.

An operational definition of instantaneous velocity at the Newtonian present requires the use of data on positions from the past. I will briefly review, though this will be familiar to some readers: Consider the present moment and a moment a finite but small time Δt earlier than the present moment. An approximation to the instantaneous velocity of a particle in the y direction is

$$((\text{y value at the present}) - (\text{y value at the moment } \Delta t \text{ before the present}))/\Delta t$$

Now make a series of such calculations with smaller and smaller values of Δt using data on the y values in the interval between the present minus Δt and the present for each calculation. Carrying out that process for the calculation of the motion of the three particles, shown in Figure 2.1, gave a graph for one of the particles shown

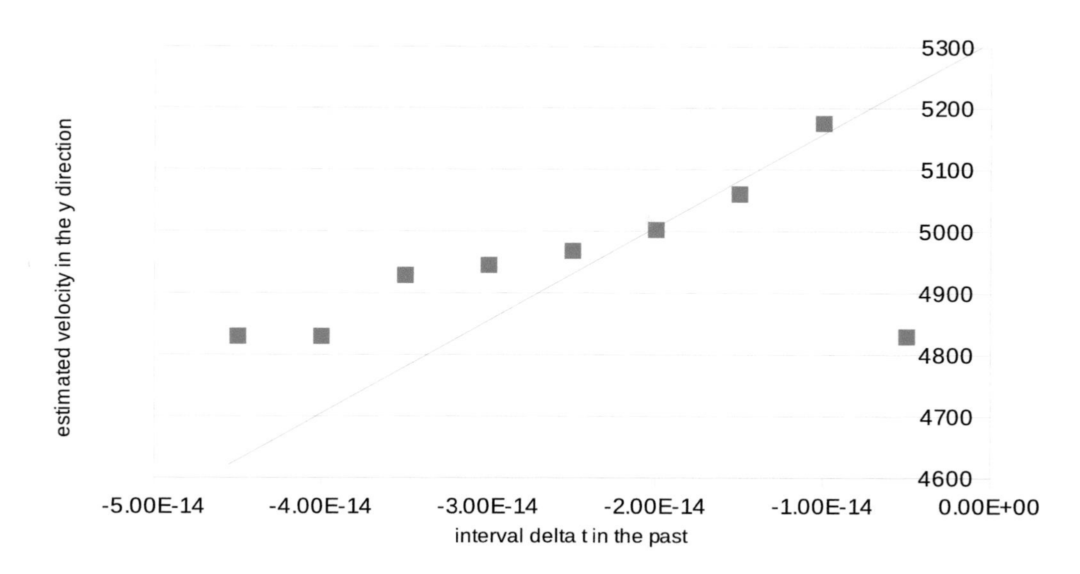

Figure 2.3 Illustration of the limiting process defining instantaneous velocity. Here some data from the simulation which gave Figure 2.1 was used. Time 0 is the present and the estimates for the velocity using positions at times farther and farther into the past are shown farther and farther to the left in the figure. The sloping straight line indicates the extrapolation process which results in an estimate of about 5300 cm/s for the instantaneous velocity at the present time 0. The units and forces chosen are the ones which are of the approximately correct size if the three objects in the simulation were atoms on the surface of a solid. The value for the estimated velocity at the shortest time interval of $.5 \times 10^{-15}$ sec was not used in the extrapolation, because the estimates become very unreliable at very short time intervals.

in Figure 2.3. If the instantaneous velocity in the y direction of the particle is well defined then it will be possible to extrapolate the resulting curve to get a value, which is the instantaneous velocity, at $\Delta t = 0$ as suggested by the line in Figure 2.3. Another example is described in detail in Appendix 2.1.

This is a limiting process, familiar to those with some knowledge of calculus, and made rigorous through the efforts of many mathematicians in the three centuries since Newton and Leibniz started using it in physics. Our main point here is that the process is impossible without use of data from the near (not immediate) past. Thus the information about the Newtonian present embodied in positions and instantaneous velocities cannot be obtained without knowledge of the near Newtonian past. In that sense the Newtonian present is not quite an infinitesimal point. The information given in the positions and instantaneous velocities of the present state is a kind of shorthand summarizing some features of the near past as well as the present instant. This is different than the human present, whose duration is limited by biophysical considerations, but, in the sense discussed, it is somewhat misleading to characterize the Newtonian present as being of infinitesimally short duration.

A lack of clarity about this point is the origin of the second of Zeno's paradoxes. Zeno's paradoxes were allegedly conceived by the philosopher Zeno in Greece more than 2500 years ago. Their immediate source is the writings of Aristotle. They have been interminably discussed by philosophers interested in questions concerning time up to the present era. This second paradox concerns an arrow in flight. There are various statements of the paradox, but in one form it states that since the arrow is seen to be in one place at a given moment it cannot be moving. This sounds a little absurd when stated in that way but the point is illustrated more clearly by considering a movie of the arrow. (Of course Zeno didn't have a movie.) If you look at just one frame of the movie, then there is no way that you can tell whether the arrow is moving or not. It is only if you also look at a frame before (and/or a frame after), the moment of interest that you can tell. Nevertheless, the movie can be used in the way cited above to get an estimate of the instantaneous velocity at the point of interest and that instantaneous velocity can be said to characterize the motion of the arrow at the instant. As before, we see that this characterization of the Newtonian state of the arrow cannot be accomplished without knowledge of the near past as well as of the instant of the Newtonian present. (In the case of the movie we could also use near future data, but if we are talking about the present, then future data is not available.)

The nature of the human present may be discussed in another way, not referring so specifically to its length, by referring to mental 'consciousness'. When one refers to the 'consciousness' of a human or animal one usually means a mental state (more specifically a state of its nervous system including the brain) which makes a kind of model or map of the organism's immediate environment 'in the present'. We do not think that an organism is fully 'conscious' if it's asleep and dreaming about a past event, under anesthesia or is otherwise not fully aware of its surroundings. The precise meaning of 'consciousness' is hotly debated and its neurological basis is not fully understood (for a comprehensive review, see [36]), so little more can be said about that aspect at this time.

The growing understanding of neurological processes provides a helpful perspective on another psychological puzzle related to the relationship of human consciousness to the present, namely the question of whether humans are moving into the future in some sense or, instead, sitting stationary in the present while time flows by them. The first option seems preferable since we have discussed the problems with making sense of any idea of 'flow' of time. At an intuitive level, most humans don't sense that they are stationary while events occur, though that impression is sometimes reported and one can find both perspectives described in common phrases such as 'I am working toward my degree' and 'The examination is coming up'. From the neurological perspective the puzzle is somewhat clarified as it becomes clearer that what we think of as a human is a series of physical events and not a stationary object. The concept of a 'human being' is a macroscopic one, in a sense to be discussed in more detail in the next chapter and is a shorthand for an immensely complicated physical system which is continually rearranging and rebuilding itself dynamically. Thus each human might be regarded as a series of events, and is different at each moment in time. If one regards a human as a series of events occurring along

Newton's timeline then it appears that the human is neither moving along the timeline nor sitting still. The human is a different set of events at each point along the line. As we have discussed with regard to clocks, this set of events makes records in its memory of other events at each moment, including clock like events (oscillations in cesium, rotation of the earth) which permit it to conceive and measure time.

2.5 THE NEWTONIAN AND THE HUMAN PAST

Looking to the past, a Newtonian certainly agrees that it cannot be altered. For a Newtonian, the future can't be altered either. What is different for human daily experience is that the past is also DISAPPEARING because of the uncertainty of personal and collective memory. In a Newtonian world, the state of everything is defined by the positions and velocities of the particles in it, and the history of the world is given by the history of how those positions and velocities change as time changes from the past up to the present. According to Newton, there is no doubt or uncertainty about it: Given the positions and velocities of all the particles in the universe at some starting point, those positions and velocities will follow one and only one path through space and time. This description, sometimes called the 'clockwork universe', was made famous by the eighteenth century French mathematician LaPlace. The past for humans in everyday life, and in most scientific and professional contexts as well, is extremely different. The human past is not known exactly and knowledge of it grows more incomplete and uncertain the farther back you attempt to look in time.

One may say that the personal past of a human extends, in duration and in precision, as far back as her or his memory and recall ability allows. In that sense, older people with good memories have more past, perhaps one should say 'accessible past' than younger people whereas older people with Alzheimer's disease have hardly any accessible past at all. Thus a full understanding of the past of humans will require an understanding of the biological mechanisms of memory.

Those biological mechanisms of memory are only partly understood but some things are known. There appear to be at least three separate processes associated with human memory, called short-term memory, long-term memory and retention. Short-term memory is used to store short, detailed pieces of information and presumably evolved so that the information could be immediately used. The storage occurs and fades quickly (over less than 30 seconds) and only a limited amount of information can be stored (three to five words, for example). For example, you may hear a telephone number and remember it long enough to write it down or dial it but then you may quickly forget it. Long-term memory requires a longer storage time, but as the name implies, it stays in memory much longer, often for many years. The amounts of information which can be stored are much larger. The limits are not fully known but I will estimate a bound later. Some understanding of the physical mechanisms of short and long-term memory formation has been gleaned from neurophysiology in recent years as I discuss later. Thirdly there is evidence that a lot more information concerning experiences than can be retrieved from long-term memory is retained in the brain but is not consciously recallable.

It has been known from ancient times that human memory can be made more accurate and that more information can be stored and recalled through effort and practice. Thus, in a sense, humans can extend their pasts by training their memories. Emphasis in education on the training of memory has declined. The first decline occurred several thousand years ago with the introduction of writing and books. Before that, the only way to retain information about the past was through human memory. Courts employed bards who retained history in memory aided by poetic rhyme, meter and musical melody. After history began to be written down, probably earliest in Babylonia, the emphasis on human memory to retain the collective human past diminished. We are now experiencing what may be a similar decrease in emphasis on human memory with the advent of the internet. It is now so easy to retrieve even the most basic kinds of information that many people have stopped trying to remember as much information as they did as little as a decade or two ago. For example, many people do not recall telephone numbers, addresses and the geographical layouts of cities in which they live to nearly the extent that they did quite recently. Though it is possible to function with this diminished store of biologically recallable information as long as the devices which electronically store biologically forgotten data continue to operate, it can be said that it diminishes the human past, in the sense we have discussed, of the people who live that way. Whether it diminishes aspects of functionality as well is not known, but it has been suggested [10].

Another aspect of the functioning of human memory has been revealed by recent psychological studies [31]. As measured in a variety of ways, it has been shown that people generally recall experiences of the past in an extremely uneven way: Experiences associated with intense feelings are remembered vividly while intervening experiences are remembered much less clearly, if at all. As a result, the duration of time between intense experiences is inaccurately remembered and often underestimated. In a sense the clock associated with memory runs faster through the less intense experiences and slower through the intense ones. One example among many in reference [31] illustrates the point: People undergoing a painful colonoscopy recorded their experience of pain minute by minute during the procedure. Then they were asked to rate how disagreeable the experience had been. Ratings of the whole experience turned out to depend only on the intensity of pain during the minutes of most intense pain and toward the end of the procedure. They did not depend on the time duration of the procedure at all. Kahneman presents evidence that longer term human memory of entire lives, both of oneself and of others, functions similarly: Lives are recalled and evaluated in terms of a few peak experiences and the end point, with little regard for the time duration of either pleasant or painful experiences in between.

Very little is known about how the brain of humans but also of other animals achieves information storage. The brain is known to be a complex network of nerve cells called neurons which can be very much longer (up to a meter) and rather wire-like compared to other cells in the body. Each neuron contains an axon, which is a fiber along which output electrical signals flow, as well as shorter fibers, called dendrites, which receive input signals from the axons of other neurons. The signals from the dendrites are passed to the axon through connections called synapses. One can

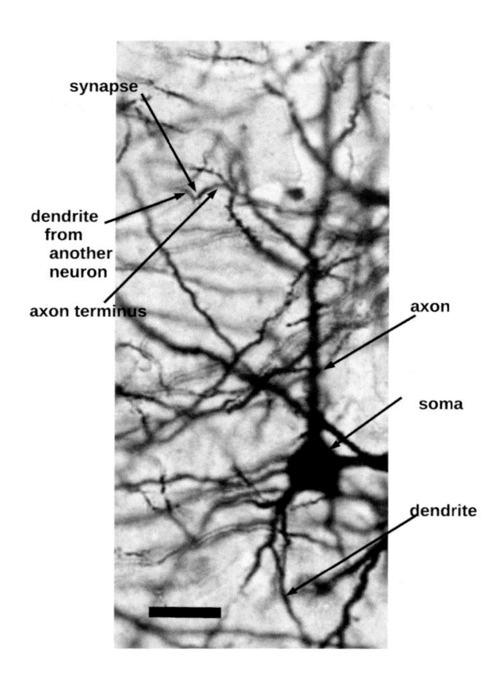

Figure 2.4 Micrograph of neurons in the human brain. The horizontal bar at the bottom of the figure is about 5×10^{-5} meters long.

regard the neuron as a device which processes inputs from the dendrites and produces electrical outputs in the axon. In that sense there is a qualitative similarity to the transistors of a computer. There are many differences, however. Notably, the number of dendrites per neuron is, on average, large (order of 10^3) whereas in electronic computers there is typically one input channel and one output channel per transistor. Neurons also operate a lot more slowly than transistors. The biochemistry by which signals are transmitted at synapses is quite well understood. A cartoon illustration of the processes known to take place appears in Figure 2.5.

Hebb [27] suggested nearly a century ago that memory is encoded in the brain by modifying the strength of synapses. The possibility that synapses can be altered in that way is called synaptic plasticity. Though not totally established, the idea been supported by recent experiments on rodents showing that the ratio of the numbers of the transmembrane proteins AMPAR and NMDAR to one another is higher in the synapses associated with neurons in memory traces, called engrams [30], in the brain. The atomic structure of the membrane protein AMPAR is shown in Figure 2.7. The function of the transmembrane proteins is to allow the signal coming down the axon to pass through the synapse to the dendrites. The synaptic signals themselves are carried by smaller molecules called neurotransmitters, indicated by dots in Figure 2.5. For example when the neurotransmitter glutamate, whose structure is indicated in Figure 2.6, is accepted by AMPAR, the transmission of a current of positive ions including Ca^{2+} is shut off.

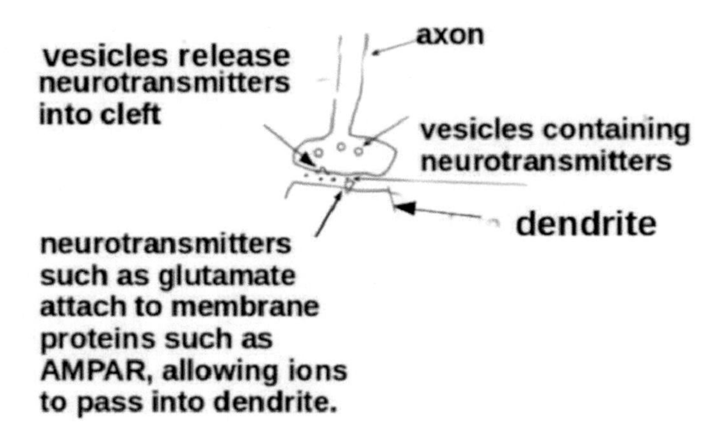

Figure 2.5 A cartoon illustrating the biochemical processes currently believed to be occurring in the synapses which connect one neuron in the nervous system to another. The end of an axon is at the top of the figure and the end of a dendrite is at the bottom. Neurotransmitters are carried in approximately spherical containers called vesicles down the axon. They attach to the membrane at the end of the axon and empty their contents into the space between the axon and the dendrite by a process called exocytosis. Control of passage of the signal to the next neuron, attached to the dendrite is achieved through the membrane proteins, including AMPAR and NMDAR, which are embedded in the dendrite surface and control whether the neurotransmitters pass the ion currents carrying the signal into the next neuron.

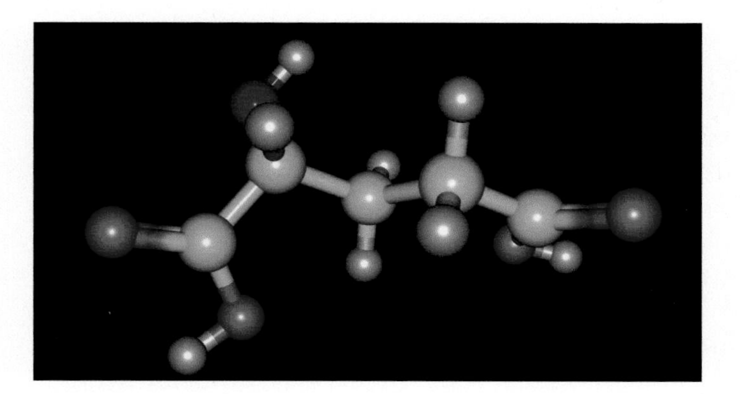

Figure 2.6 The atomic structure of the neurotransmitter glutamate. Light blue spheres represent carbon atoms, red spheres represent oxygen, gray spheres represent hydrogen and the dark sphere represents a nitrogen atom. Such pictures accurately represent the relative atomic positions of the atomic nuclei in the molecule, but they do not indicate in any realistic way what the molecule 'looks like'.

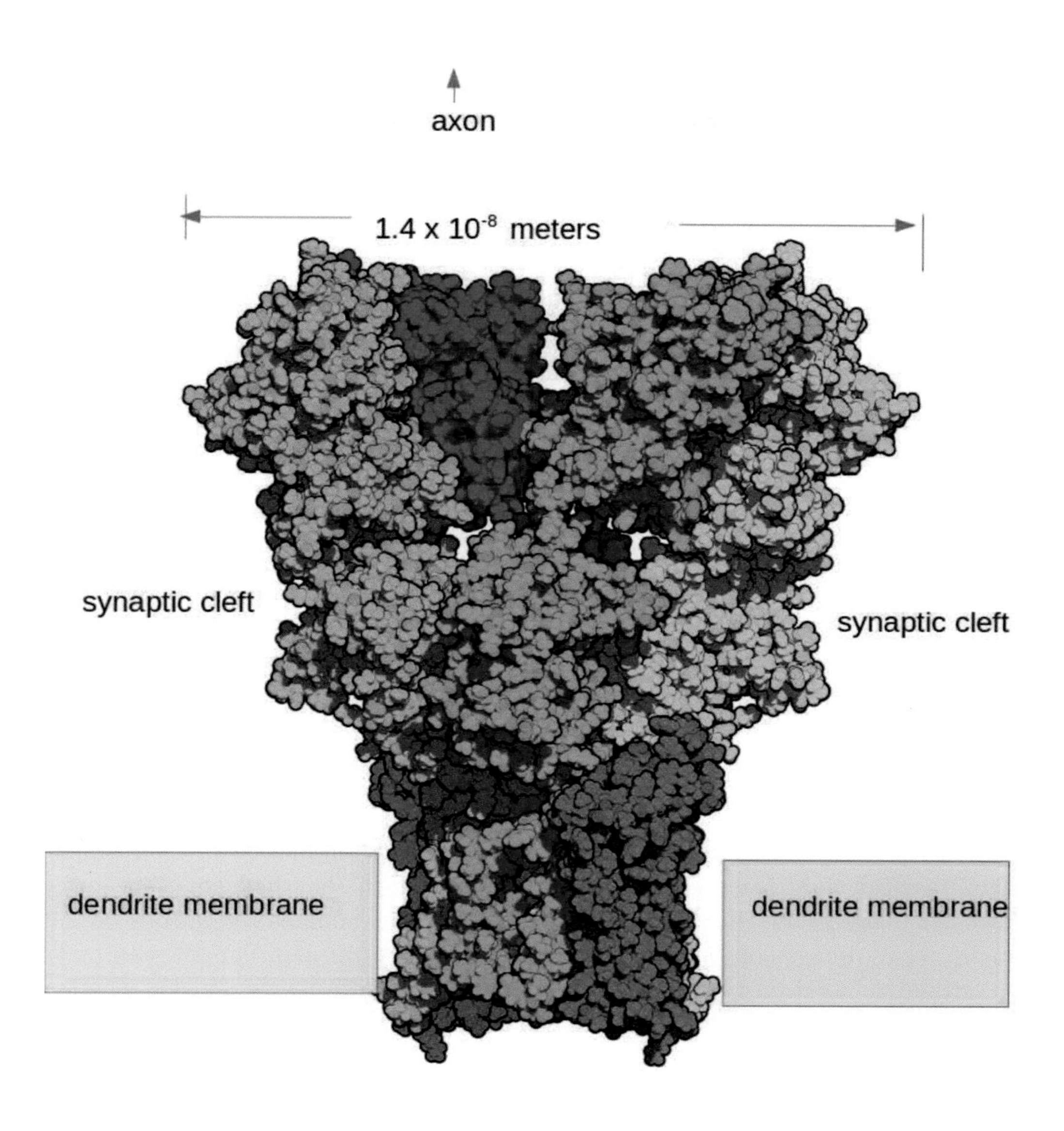

Figure 2.7 The atomic structure of the membrane protein AMPAR, which provides a connection at dendrite surfaces allowing or shutting off passage of positively charged potassium and sodium ion currents (like a switch) to the dendrite of the next neuron when activated by the neurotransmitter glutamate. The neurotransmitter glutamate binds to a region of the protein inside the cleft and near the dendrite membrane where it induces a conformational change in the transmembrane domain shutting off the current [41]. There is a passage running through the protein, vertically in the picture but not visible, through which the currents pass when the AMPAR is 'open'

Some understanding of the way in which different parts of human and rodent brains store and recall memories has recently emerged from neurobiology. The process of collection of sensory data and assembling it into a coherent picture is believed to occur in the part of the brain known as the hippocampus. Storage as long-term memory is then affected by transfer of the memory to the prefrontal cortex [39]. It is suggested that synapses are only biochemically altered (synaptic plasticity) during the formation of long-term memory in the prefrontal cortex. Some evidence is reported suggesting that elements of the assembled picture are stored separately in the frontal cortex and then reassembled in the hippocampus upon recall [40]. On shorter time scales, detailed evidence has been reported indicating that bats carry a three-dimensional respresentation of their immediate past, present and a prediction of their immediate future positions in their hippocampus as they fly through caves [16].

One can quite easily get a little more insight from the size of the network in the brain. There are reported to be roughly 10^{12} neurons and 10^{15} synapses in the human brain meaning that there are about 1000 synapses per neuron. As just described, long-term memory is reported [20], [30] to be associated with chemical changes in the synapses. If one postulates that each synapse stores a bit (equivalent to a 0 or a 1), then the storage capacity of the brain is of the order of 10^{14} (8 bit) bytes or 10^5 gigabytes. That is a lot more potential storage capacity than most computers have, if all the synapses could be used for memory. For example, the number of synapses is about 4 orders of magnitude larger than the number of transistors in the CPU of a typical laptop (Intel Atom N450) in 2010. However, human information processing is a lot slower than electronic information processing: Clock speed for the same CPU is 1.6×10^9 s^{-1} operations per second whereas the human clock speed is of the order of 10^3 s^{-1}.

On the basis of experiments in which, after physical stimulation of their brains, humans recalled vivid pictures of past experiences which they had not consciously recalled for many years, there has been speculation that humans retain, but do not normally recall, images of virtually all their experiences through life. We can make some estimates based on the numbers in the last paragraph to get an idea whether such total retention would be physically possible. Suppose that each experience of total recall contains an amount of information equivalent to that in a 216×216 pixel video image and that each pixel is described by an 8-bit byte. If such an image were stored every second for 30 years (supposing that acute visual perception only goes on about 1/3 of the time) then the amount of required storage would be about 3×10^{13} bits which is about two orders of magnitude less than reported number of synapses. Thus, total retention might not be totally impossible but it would be expected to begin to be more difficult toward the end of a long human life. Psychologists speculate that much retained information is not accessible to conscious recall to permit attention to be focused on a smaller amount of information useful for guiding actions beneficial to survival and reproduction.

2.6 THE COLLECTIVE HUMAN PAST

There is also a collective social human apparatus for retaining information about the past, including most notably libraries and machine readable databases though cultural practices in commerce, family life and religions also play that role. Libraries are to some extent morphing into machine readable data bases. The magnitude of information storage is larger than in an individual human brain and information is retained on time scales longer than a human lifetime, though not indefinitely. Currently technological developments have permitted huge increases in the data stored in this way as well as in methods of retrieval. The impact of these revolutionary developments is unclear. It appears that much of the data stored is of low quality and the expected lifetime of the digital records is short. Magnetic memories must be rewritten every few years. The estimated lifetime of optically written compact discs is about 40 years. Microfilm records, which are in decreasing use, have an estimated lifetime of centuries. Documents printed on paper have estimated lifetimes which depend on the quality of the paper used. Information on high quality low acid paper will survive for centuries, but from the 1920's to quite recently most paper documents were printed on high acid paper which deteriorates within 50 years. The latter process is currently destroying millions of volumes of printed information stored in libraries throughout the world and very little is being done to retrieve and save the information which they contain. One can safely say that rapid changes in the technology of information processing and storage are significantly altering the meaning and content of the collective human past and may be effectively both enlarging it in volume and shortening it in longevity significantly. Many issues arise as collective memory is increasingly digital and expands spectacularly in volume while the records grow much less durable. The volume growth is illustrated in Figure 2.8, from a study of global information storage capacity in the last decade. The estimates indicate that the amount of digital storage worldwide was about 2×10^{22} bits in 2021. (2500 exabytes. An exabyte is 10^{18} bytes.) That's about 25 million times the number of synapses in an individual human brain. The changes in volume of stored information have been doubling approximately every 3 or 4 years. At the same time, the expected durability or lifetime of those information resources are getting shorter and shorter as the dominant technological means of storage has moved from paper (books and journals) to one digital storage medium after another.

The natural record of the past includes geological data providing information to 4.5 billion years ago and astronomical data to 13.8 billion years ago, though the quality and abundance of the natural record decreases as one goes further back in time. For human consciousness, the natural record has to be recorded and interpreted. With new means of observation and collection, enormous amounts of new data on this natural record are being collected, particularly in astronomy, earth sciences and molecular biology. Interpretation requires organization and, in most cases, mathematical modeling of the data and that is proving a formidable challenge in many scientific fields. A further issue associated with the collective human memory arises in sociology and related sciences which have a new trove of information on human activities through records of internet activity. Much of this data is not universally

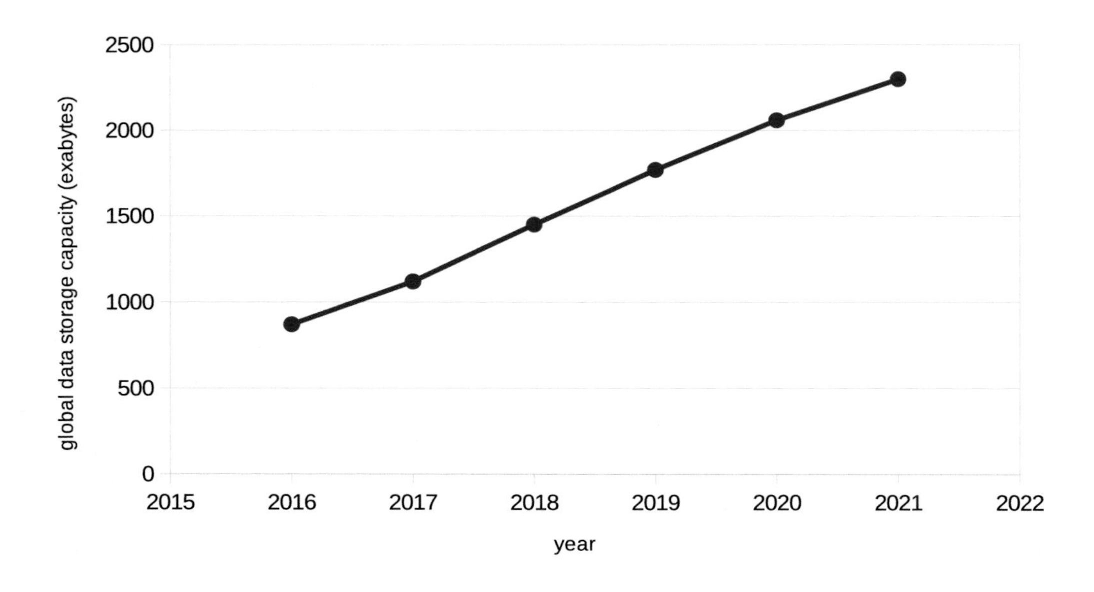

Figure 2.8 Estimates of the volume of data stored by humans collectively during the period 2016 to 2021. The growth rate has slowed significantly during this period and is a lot lower than in the immediately preceding decades. Drawn from data in Cisco Systems Statista estimates, Statista 2021.

accessible, even for those with the needed electronic capabilities, and there are indications that accessibility is decreasing even as the amount of information stored continues to increase. Ongoing discussions of how to use and collect this data raise issues of privacy and the ethics of legitimate use. For example, medical records provide useful information for treating and controlling the spread of disease, but using that data raises questions of how to protect the anonymity of the patients whose data is being used. Tracing people's movements using GPS data can be useful to improve traffic and disease control as well as other aspects of urban planning. But again, the applications can be in violation of generally accepted privacy protections. There are obvious potential applications in business (targeted advertising) and government. Election campaigns and fundraising informed by analysis of massive amounts of demographic data analysed with new artificial intelligence techniques can potentially manipulate governmental outcomes while the targeted populations are unaware of the manipulations taking place. Such manipulations have occurred both within nations and internationally, sometimes with purely malicious intent. Thus the human record of the near human past has recently become enormously deeper and the potential uses of this more detailed information threaten to destabilize many human institutions. At the same time the new data on the human past offers unprecedented opportunities for improvement of the human condition through better transportation, housing, medical and economic planning and climate, provided that approriate rules can be formulated and enforced.

2.7 THE NEWTONIAN AND THE HUMAN FUTURE

How about the future? Newtonian physics makes an EXACT prediction about the future as well. Given the present positions and velocities of all the particles at the present, their positions and velocities (and therefore, within a Newtonian world, the state of the world) are known in principle for all future times. (The methods by which such Newtonian predictions can be made are discussed in some more detail in Appendix 2.1.) Again for humans in practical life, hardly anything in the future, except perhaps the trajectories of some beautifully designed space probes, is like that. Predictions, when they can be made at all, are short term, uncertain, and often wrong. Thus, for the future, the contrast between the human and the Newtonian view looks stark because the Newtonian thinks it is fixed and modern humans mostly think it is 'open' and that they can do something to affect it, though in fact they are not good at predicting it. These 'folk hypotheses' about the future: openness, unpredictability and responsiveness to human will are hard to formulate in a way which is free of internal contradiction. That is related to the long philosophical dispute about free will.

To see that the Newtonian and human perspectives are not quite as different as they might seem at first, I have to bring up some relatively recently discovered facts about the mathematical nature of complicated Newtonian systems. Such systems often have the property of being 'chaotic' in the following technical sense: It is true that if all the positions and velocities of all the particles are known EXACTLY in the present, then in Newtonian physics the subsequent trajectory is determined, also exactly. However in chaotic Newtonian systems, TINY changes in the present positions and velocities can and do result in HUGE changes in the future trajectory. A very notable example of considerable social significance is in weather prediction and climate science. The physics of the climate system is quite accurately Newtonian, but the equations which describe it are intrinsically chaotic. Predictive capability is correspondingly difficult to achieve. A popularly accessible account of this and related discoveries about Newtonian systems can be found in [23].

The chaotic nature of the underlying physics is at least part of the reason that humans experience the future as 'open'. Daily experience teaches that small adjustments can make big differences in results but also that prediction is very difficult, 'especially when it involves the future' (a joke attributed to many authors but probably first published by Danish politician Karl Kristian Steinecke [56]). Thus though the future may very well be almost totally determined by the present (and would be totally determined in a Newtonian world); in practice, humans know too little about the present to make accurate predictions of what that future will be and therefore perceive it as 'open' and/or uncertain.

The collective human future is documented in a vast trove of human documents constituting plans and agreements for actions to be taken in the future. These include contracts to buy and sell, mortgages, treaties, trade agreements wills and bequests, architectural and development plans and educational curricula to mention only a few. These data potentially provide opportunities to gain insight into the nature and extent of the collective human future, though that has apparently not been an object of much detailed study. One relatively obvious observation is that the time scale to which such documents refer very rarely exceeds expected average longevity at the time they were written, indicating that though humans often express a concern for the farther future involving their descendants, they rarely act on that concern. Typical mortgages, for example, are written for thirty year maturity, arising from a calculation of the expected time for the occupants of the mortgaged property to raise a family and retire or die. In education there are indications that the average time spent in formal education may be a quite stable fraction of the life expectancy in the society. For example, in the US, the average life expectancy in 1910 was 48.4 years and the average time an individual spent in formal education was 8.1 years, whereas in 1970 the corresponding numbers were 70.8 years and 12.2 years. The average fraction of a lifetime spent in formal education was close to was 17% in 1910 and 18% in 1970 despite a 38% increase in male longevity. Though many factors could be at play, it appears that the expected lifetime could be playing a role in determining the amount of schooling required. Similar regularities appear in planning documents: Terms of office for governmental positions rarely exceed 5 years in democratically governed countries and in authoritarian ones, economic planning documents cover a similar span of future time. Bond and futures markets supply a rich source of data for determining the main drivers of future human planning and could be used to test the hypothesis that human longevity is a determining factor. A quantitative indicator of the time scales which dominate the collective human future is the 'discount rate' which describes the amount by which a future benefit is discounted relative to an immediate one in evaluating the worth of an investment. These annual rates are typically around 5% implying an approximately exponential decrease in the value of a dollar gained n years into the future by investment of a dollar invested now of $e^{-n/20}$. ($e = 2.717....$ This amounts to a future 'half-life' of a dollar invested now at about 14 years.) This sets a time scale of the order of 20 years. If life expectancy sets the time scale of human planning, it could help to account for the difficulty that humans are having in dealing successfully with long-term global threats such as climate change, nuclear weapons and nuclear waste disposal, pandemics and overpopulation. Though the threats are very real, their worst consequences are projected to be in the future beyond expected human lifetimes and human institutions and biologically inherited intuitions are ill equipped to deal with such long-term planning. Perhaps hopefully, there are a few exceptions, such as the European cathedrals and the US constitution, which represent partially successful human planning well beyond expected human lifetimes.

2.8 THE REALITY OF THE PAST AND FUTURE

A modern school of 'presentist' philosophers on the nature of time contends that only the present is 'real' and the past and the future are not real, but somehow, in varying senses, mere memory or imagining. That was also Augustine's view as discussed earlier. If one takes the Figure 2.2 above and expands the uncertainty in the past and future directions to be very wide, then one ends up with a perspective of this sort somewhat as indicated in Figure 2.9.

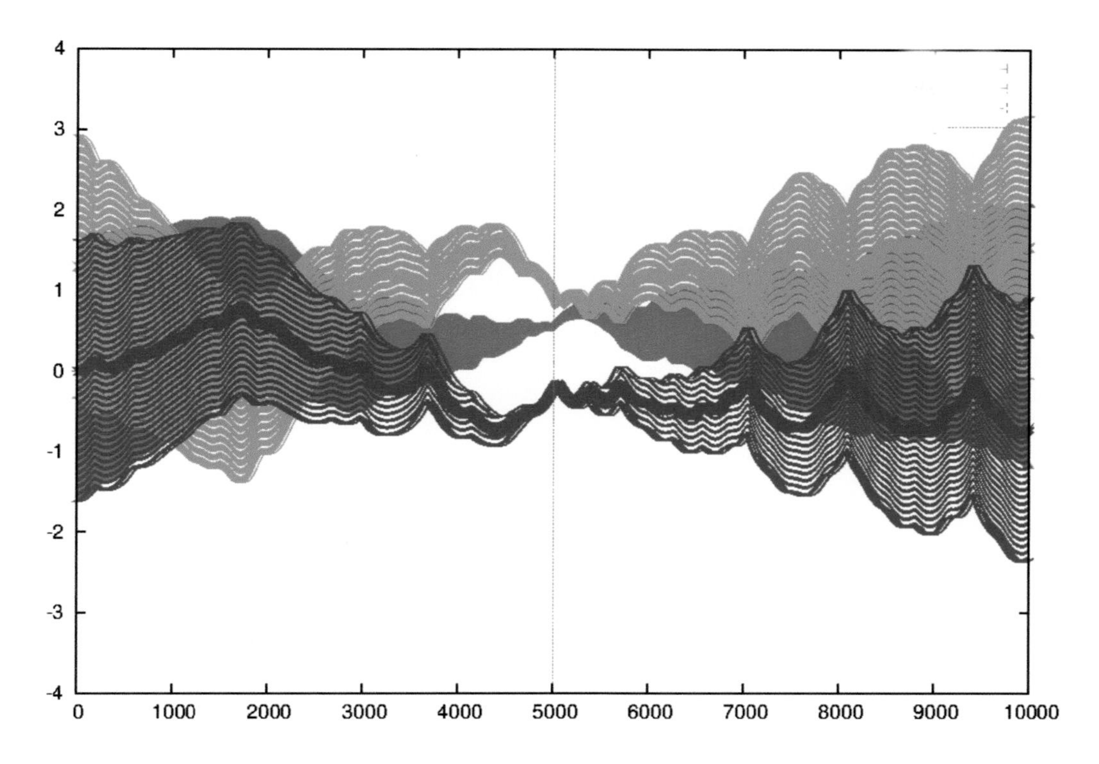

Figure 2.9 The data of Figure 2.2 in which the trajectories were assumed to become more and more uncertain as time progressed into the future and back into the past, suggesting the human perspective.

On the other hand one finds a philosophical school, and also most physicists, favoring a 'block universe' perspective in which all the events of past, present and future are equally real. That perspective is more compatible with the relativistic nature of space and time discovered empirically in the early 20th century and formulated theoretically by Einstein as discussed in Chapter 5. To start consulting the recent academic philosophical literature on the philosophical debate, see [58]. We will see that presentism has serious difficulties accounting for the results of the special theory of relativity in Chapter 5. The 'block universe' or 'eternalist' view that all the events in the past present and future are real is close to the Newtonian picture and also to a view gleaned from a more modern general relativistic view, but not like a perspective gained from statistical physics and thermodynamics, as I will discuss later. The 'block universe' perspective is like narrowing the width of the uncertainty cones in the future and past directions to have zero widths. We will see later that the picture of the human perspective as increasingly clouded (more uncertain) as one looks farther in the future and past directions, though for apparently different reasons in the two cases, can be stated a little more formally in terms of concepts of entropy and related to other aspects of the question which arise from the physics of heat.

Another intriguing aspect arises from the mathematics of Newtonian physics. It was proven by the French mathematician Henri Poincare in the late 19th century that under rather general conditions in a CLOSED system of Newtonian particles, the system will eventually return to EXACTLY the same state in which it started [47]. This is very similar to the recurrence said to be believed by the Stoics centuries ago as described in the passage quoted earlier. However estimates for the time for the universe as a whole to return to its initial state if it were a Newtonian system are immensely large and greatly exceed the present age of the universe [44]. Three comments: 1) The universe is not Newtonian, 2) probably isn't closed and 3), we would be in all likelihood be unable to see either back or forward to the recurrent state even if it were to occur. With regard to point 1) there are versions of Poincare's theorem which take account of quantum mechanics and relativity. Concerning 2) currently cosmologists believe that the universe is not closed but that issue has been in play quite recently and is probably not settled. 3) implies that recurrence, if it did occur, would have no practical human effect.

Recurrence would effectively obliterate any global distinction between the past and the future because, as one increases time into the future, one eventually arrives at the series of events which are conventionally labeled the past (Figure 2.10). (A relativistic version of such a model also exists [42].) Some of the paradoxes which have been suggested to be associated with this situation are not hard to resolve. For example, it does not imply that there is any possibility of killing or otherwise rendering one's grandmother infertile: In a recurrence EVERYTHING, including any individual organism such as oneself, would be and would behave identically in the newly reentered 'past' as it did in a previous cycle. Thus, in that reentered 'past' the grandchildren of one's grandmother would not yet exist and would pose her no threat. Science fiction speculations about 'time travel' usually assume that everthing about the past EXCEPT the traveling individual reverts to its previous state. Even if the speculative ideas about recurrence discussed here were to be realized in the real universe, such an assumption about exceptions seems totally unrealistic. It should be emphasized that even if there were evidence that we lived in a recurrent universe (and there is no such evidence), the time scales for recurrence would be almost inconceivably long and would have absolutely no practical effect on any human perspective.

A related set of questions concerns causality. Two events are said to be causally related if the causal event always precedes the caused event in time and if the caused event never occurs unless the causal event occurred first. This definition already poses some issues, as for example, in the case of a cock that always crows just before sunrise. However, leaving those issues aside and assuming that a meaningful definition of causality can be formulated which relates earlier to later events, one sees that the following additional issues arise: In a Newtonian universe, whether closed or not, it is just as reasonable to regard the future as determining the past as vice versa, so the entire notion of causality is subject to question. In fact this issue arises quite explicitly in modern mathematical formulations of the dynamical behavior of both Newtonian and quantum mechanical complex systems: Solutions to the Newtonian (or quantum) equations appear in which the future determines the present. Those solutions are routinely termed 'acausal' and usually ignored without much comment by the professionals, though there is nothing in the mathematical formulation which justifies that procedure. Further, in a recurrent universe, the past cannot be globally

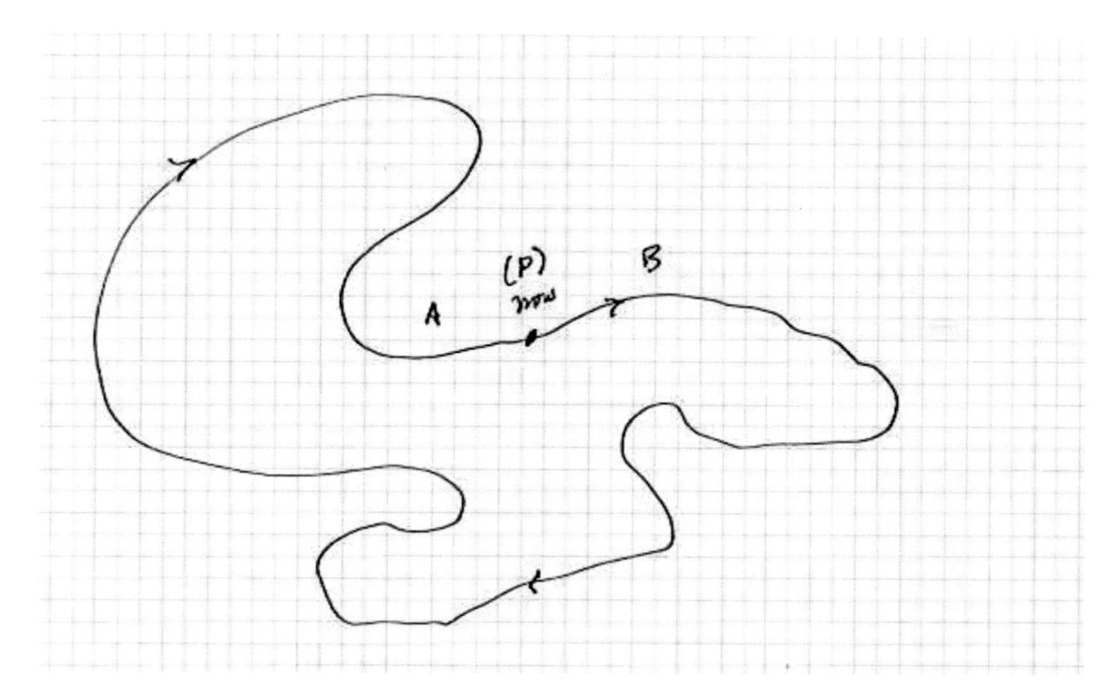

Figure 2.10 Schematic sketch of the path of a series of events in the space of particle positions in a closed Newtonian universe. If one is at the point labeled 'now', then one conventionally regards events in the region B as 'future' and events in the region 'A' as 'past' events. But it is easy to see that when the loop is closed, as sketched, events A will occur again as time evolves so they may be regarded as both in the future and in the past of 'now'. Similarly, events B have occurred before the present as well as being expected to come after it so that the events B are also both in the future and in the past.

distinguished from the future so the question reduces to whether the causal relationships run one way or the other around the loop schematically illustrated in Figure 2.10. For Newtonian physics, the answer is 'either way'. With regard to the human perspective, the issue of causality is somewhat different but the conclusions are similarly ambivalent: Some philosophers and psychologists contend that human actions are caused by a human's genetic and cultural past ('pushed from behind' [52]), while others observe the effects of the projected future on human actions ('drawn from in front', 'executive function' [7]).

CHAPTER 3

Thermodynamics, Irreversibility and Time

"The laws of thermodynamics, as empirically determined, express the approximate and probable behavior of systems of a great number of particles, or, more precisely, they express the laws of mechanics for such systems as they appear to beings who have not the fineness of perception to enable them to appreciate quantities of the order of magnitude of those which relate to single particles, and who cannot repeat their experiments often enough to obtain any but the most probable results." J. Willard Gibbs, "Elementary Principles in Statististical Mechanics" p. 8, 1902.

3.1 INTRODUCTION

To understand if and/or how the past is different from the future, the science of heat, commonly called thermodynamics, supplemented by the related discipline of statistical mechanics turns out to be relevant. Thermodynamics arose in the 19th century, significantly from efforts to understand the operations of then new technologies such as the steam engine at a quantitative level. From extensive experiments on systems of gases and liquids, later extended to other states of matter consisting of very large numbers of atoms and molecules, scientists formulated laws which appeared to never be violated for such systems. The most important of these were the first and second laws of thermodynamics.

The first law of thermodynamics stated that the total energy of any thermally isolated system always remained the same. This first law is completely consistent with a description of the system using Newtonian mechanics (and also by quantum mechanics as we will discuss later). The second law of thermodynamics stated that a quantity called entropy, which could be calculated by a well-defined procedure for such systems, would always INCREASE, or, in exceptional cases, remain unchanged in time regardless of what processes the material system underwent.

Though the second law of thermodynamics did not seem to meet with experimental exceptions, there were serious difficulties in reconciling it with a mechanical

DOI: 10.1201/9781003037125-3

description of the materials systems involved using Newtonian dynamics. The difficulties were of two kinds. First, though there was a MACROSCOPIC prescription for calculating the entropy in terms of heat transfer and measured temperatures, its meaning at the molecular level was not clear. It turns out to be a good idea to be as clear as possible about the meaning of the word 'macroscopic' here. The word 'macroscopic' was coined to contrast with 'microscopic' and the latter was supposed to mean 'at the atomic level'. Actually, when one speaks of a 'microscopic' description in this context, one does not mean 'at the level attained by an optical microscope' because an optical microscopic does not permit one to see atoms. But if 'macroscopic' just means 'not microscopic' in the sense of 'not at the atomic level' then specifying that the description is 'macroscopic' does not tell you very much. It could mean that you are specifying the properties of the system of interest at a level of detail associated with properties averaged over millimeters, meters, kilometers or light years. In the original practice of the science of thermodynamics the averaging in a 'macroscopic' description was done on the scale of the dimensions of a container (often a cylinder with a piston) of the fluid (often steam) of the container of interest and was of the order of centimeters to meters.

The question of the level of description is sometimes called the question of the level of 'coarse graining' which is used to define the macroscopic description. As long as this coarse graining is at a scale longer than the distance between atoms, the resulting entropy is observed to increase with time during most of the time. However the level of coarse graining does make a difference: The magnitude of the entropy is quite dramatically different depending on the level of coarse graining, and exceptions to the rule that it increases with time are more frequent when the coarse graining length is small. To illustrate this effect of coarse graining on the value of entropy, I provide an example below and another in Appendix 3.1. With a given level of coarse graining, the difficulty in defining the meaning of entropy in terms of the atomic description has been largely surmounted.

The second difficulty is that the growth in the amount of entropy in any isolated system with time implies that the thermodynamic theory of matter is NOT TIME REVERSIBLE. What that means is that, if you took a movie of any process obeying the laws of thermodynamics and ran it backward, you would observe a sequence of events which, according to thermodynamics, and in fact according in most cases to everyday experience, is never observed. Examples from everyday life abound: Drop an egg on the floor and observe a process quite commonly observed. Run a movie of the events backward and observe the mess on the floor reassemble itself into an egg in your hand. The latter sequence is never observed in nature.

That seems clear enough until one tries to reconcile it with the description of the egg in terms of the molecules of which it is undoubtedly made. It is known that those molecules are obeying Newtonian equations of motion to an excellent approximation. For every sequence of events allowed by the Newtonian equations of motion, the same sequence running in the opposite direction in time is also allowed. (Newtonian mechanics is TIME REVERSIBLE). So the events you observe in the backward running movie are entirely possible according to Newtonian mechanics, which should work well here, yet they are forbidden by the second law of thermodynamics and are never observed. What is going on?

Clearly this has something to do with the difference between the past and the future: The second law of thermodynamics states that the future is different from the past because the matter of the universe will have more entropy than that matter had in the past. (All these statements can be generalized to include electromagnetic radiation such as light and radio waves, which carry energy and exert forces on matter but do not have mass. So the last sentence before the parenthesis is true if you substitute 'matter and electromagnetic radiation' for 'matter'.) In fact, quite a number of authors have chosen to regard this difference in thermodynamic properties as defining or determining an 'arrow of time' which establishes a direction for time pointing from the past toward the future.

3.2 THE MICROSCOPIC DEFINITION OF ENTROPY

To try to understand the situation a little better, we return to the question of how entropy might be defined at the level of atoms and molecules (I will sometimes say 'at a microscopic level'.) Historically, the amount of entropy in a system was first calculated by use of measurements of the heat flowing in or out of it and the temperature. This worked well to establish the law of its increase but the molecular meaning of the resulting quantity was unclear. Boltzmann is credited with first suggesting that the entropy was measuring the number of ways that the constituent molecules could be arranged and still give a system with the same macroscopic description (by specifying temperature and pressure, for example). A key point is that temperature and pressure, do not fully specify the Newtonian state of the system, which I have emphasized requires the specification of the positions and velocities of all the atoms in the system. That's a LOT more data than just the temperature and pressure. In a macroscopic system like a piston full of steam vapor there are something like 10^{24} atoms so to specify the Newtonian state exactly in molecular terms would require around 6×10^{24} numbers, not just two. Of all the huge number of possible Newtonian states, many would not give the observed temperature and pressure of the vapor but a very large number of the states would. So Boltzmann said, take the logarithm of the number of Newtonian states associated with a given thermodynamic state (specified, for example, by pressure and temperature) and that will give you the entropy. (There are good technical reasons for taking the logarithm but we will not go into them here. One obvious, but not sufficient, reason is that the number of states is going to be huge and taking its logarithm makes it more manageable.)

We may summarize this idea in an equation, in which I follow tradition and denote the entropy by the symbol S:

$$\text{(entropy S of system with some coarse-grained description)}/k =$$

$$\ln(\text{number of Newtonian states associated with that coarse-grained description})$$

Again following tradition, I have included a constant k which is of no importance to this discussion and have denoted the logarithm by the symbol ln which refers to taking the logarithm to a particular 'base'. There are also some technical issues associated with how you count Newtonian states which have not been discussed. None of these points is essential to the discussion which follows. What IS important is that this definition of the entropy of a system DEPENDS ON THE LEVEL OF DETAIL IN ITS DESCRIPTION (ie the coarse graining). That means that identically the same system in identically the same state can be determined to have different values of entropy depending on how precisely you describe it. If I have a system of molecules obeying Newtonian mechanics, then if I tell you positions and velocities (and masses) of all the particles, then there is only one such possible microscopic state and the entropy using that level of description is $S/k = \ln(1) = 0$. And if I exactly solve Newton's equations for the subsequent behavior of that system and continue to describe it with the same level of detail, the entropy will remain exactly zero. On the other hand if I only describe the same system by the number of molecules and the temperature and pressure, then there is a huge number of Newtonian states consistent with that limited description and the entropy will be much larger than zero. To further emphasize the point that the entropy depends on the level of detail in the description I consider one simple example here and some others involving cards in Appendix 3.1.

3.3 HOW THE ENTROPY GROWS IN TIME

In the example discussed next, we envision a toy universe in which the most detailed level of description possible is the specification of the color and number coming up on each of two dice of different colors, corresponding to the 'microscopic' description and the other descriptions described correspond to differently coarse-grained 'macroscopic' descriptions. Consider a roll of the pair of dice. Suppose one of the dice is red and the other is blue. Consider the following four state specifications corresponding to increasingly coarse grainings. The first specification a) is 'microscopic' in this toy example: it tells you as much as you can possibly know about the given objects. The other three specifications (b),(c) and (d) are 'macroscopic' descriptions of increasing coarseness:

a) which numbers are up on red and on blue

b) the pair of numbers, but no specification of which goes on red and which on blue

c) the sum N of the numbers which come up.

d) Just the statement that there are two dice of different colors with one of the six numbers facing up on each.

What is the entropy in each case? (I encourage you to think about it before looking at the answer on the next page.)

TABLE 3.1 Table for case c.

N=	2	3	4	5	6	7	8	9	10	11	12
cases	11	12	13	14	15	16	26	36	46	56	66
		21	22	23	24	25	35	45	55	65	
+			31	32	33	34	44	54	64		
				41	42	43	53	63			
					51	52	62				
						61					
cases	1	2	3	4	5	6	5	4	3	2	1
S/k	0	ln 2	ln 3	ln 4	ln 5	ln 6	ln 5	ln 4	ln 3	ln 2	0

a) $\ln(1) = 0$ for all cases
b) for pairs 11, 22, 33, 44, 55, 66 $S/k = \ln(1) = 0$ for all others 12, 13, 14, 23, 24,... etc $S/k = \ln(2)$
c) See Table 3.1.
d) ln(6 possible numbers on the first times 6 possible numbers on the second) $= \ln(36)$

Now we can start to get an idea why the entropy might increase in time at least under certain conditions. Let's start with the last example and suppose that you are using the 'coarse graining' c) (which is the common one used in scoring dice). You have just rolled a 2 ('snake eyes', two 1's). The entropy, as we have defined it, is 0 as listed in the table, because there is only 1 way to get $N = 2$. So you are starting from a low entropy state. Now you roll again. You might try to roll just as before because you want to get $N = 2$ again, but it is extremely likely that you will fail, even though the Newtonian trajectory of the dice is completely determined, because tiny changes in the way you throw the dice lead to completely different scores. The most likely result of the second roll is $N = 7$ because there are 6 ways out of the 36 total possible outcomes to get that, whereas there is only one chance in 36 of getting a 2 again. All the other possible scores except $N = 12$ are more likely, and they all have higher entropy than snake eyes again. Thus the entropy is very likely to increase on the next roll. On the other hand if you start with a high entropy score such as 7, there is a good probability of getting 7 again, but it is not a certainty. The entropy could go down.

Notice several features of the last example: The physics of rolling dice is all Newtonian: Given an exact description of the initial velocities and positions of the components of the dice, the trajectory and outcome of a toss are exactly determined. But the trajectories are very sensitive to small changes in those initial conditions and that makes it impossible to control the final state. As a result, in most cases the result of a toss can only be said to be expected to be most likely for the scores which have the most ways of being realized, ie which have the highest entropy. We see that the approximate realization of the second law in this toy model depends on 1) a coarse-grained definition of the 'macrostate' here defined by N and 2) the sensitivity of the Newtonian dynamics to the starting conditions (termed, a little loosely, the 'chaotic' nature of the dynamics). The fact that the entropy as defined here could go down if you start from a high entropy state is also characteristic of the second law: There can be fluctuations of the entropy, even for more complex systems, in which the entropy momentarily falls. However, in many systems, those fluctuations can be shown to get smaller and less likely as the systems get bigger.

Similarly, consider the dropped egg. The molecules in the egg are so highly organized that if you weren't going to break it and eat it they were going to arrange themselves into a baby chicken. Though there are definitely a lot of arrangements of the molecules in the egg which would lead to that result, the number is much much smaller than the number of arrangements that would not. So the entropy of the egg is relatively low. You may think of the Newtonian state of the molecules in the egg as being quite precisely aimed at the production of a chicken. Now you drop the egg. You perturb the Newtonian state of the molecules. In some sense you don't perturb it by a huge amount, but now we invoke the chaotic property of the Newtonian dynamics. The perturbation is sufficient to completely spoil the aim of the original Newtonian state and the Newtonian equations predict that depending on exactly how it drops, there is a huge number of very different ways in which it will be predicted to end up on the floor. If our description of the final state is just 'egg broken on floor' then there are many more molecular ways of realizing it than there were of the egg before it dropped so the entropy has increased. The reason that entropy increased is seen to be associated with 1) the fact that the Newtonian equations are chaotic, so that small perturbations of the initial molecular state can send the egg into one of many very different but macroscopically equivalent states thus increasing the entropy and 2) the system started out in a low entropy state.

What about running the movie backwards? Well what if we tried it in the real world? That is what if we tried somehow to put the egg back together (as in the nursery rhyme)? By one means or another we would have to direct the molecules back from their highly disorganized form into the original organized form. Given the chaotic nature of the Newtonian equations, this would require exquisite fine tuning of the aiming of each molecule or set of molecules. Without the help of a very talented microbiologist, it would certainly be impossible and would not occur spontaneously. (Another problem with time reversal in this example is that the egg undoubtedly perturbed the molecules in the floor when it hit so one would have to somehow reverse the velocities of those molecules as well. Proper account of the such effects of the environment of a system of interest on its state are a major part of the quantitative formulation of thermodynamics but they will not concern us very much here.)

It may help to visualize the process by which entropy grows a little more abstractly, with a picture. In the Figure 3.1, each point in the space is intended to represent a Newtonian state (for example of the egg). On the left, the small solid oval represents a region of this space of states which can describe an unbroken egg. When you break the egg, the Newtonian state of the egg, say A, goes (deterministically and uniquely) to a state, labeled A', in the larger oval on the right. That oval encompasses all the Newtonian states consistent with the macroscopic description 'broken egg'. The fact that the entropy of the broken egg is larger than the entropy of the unbroken egg is indicated in the picture by the fact that the small oval on the left is smaller than the oval on the right. The reason the oval on the right is larger is because, if you start with another intact egg, say at point B on the left, then the chaotic nature of Newtonian dynamics tells you that it will end up someplace very different, say at B', which is far from A'. Thus the macroscopic description of the broken egg will be consistent with a larger number of Newtonian states, as indicated by the larger oval. (That large oval also contains a lot of Newtonian states which are not the end points of trajectories starting in the small oval but which are also within the set of states described macroscopically as 'broken egg on the floor'.) Now if you could precisely turn the velocities of all the molecules in the broken egg state A' at the end of the indicated path around so that they went in the exactly opposite direction to the direction in which they were moving then the path would be exactly retraced back to A in the small oval and the egg would be restored because Newtonian physics is time reversible. (Actually, because the some of the molecules interacted with molecules in the floor, you would have to invert the velocities of those molecules too for this to work. The same would be true of the B-B' path.) But if all you knew was that the egg was dropped on the floor and not the exact locations and velocities of all the molecules at A', then it is extremely likely, because there are so many Newtonian states in the right oval, that when you try to reverse the process and restore the egg, you will miss and make a time reversed state that might be near, but is not exactly, the time reverse of A', say at C'. But most of the states in the right oval are not like A' and B' which are the exact result of the time evolution of an egg being broken. So time reversal of C' is overwhelmingly likely not to take you back to the unbroken egg. The chaotic nature of Newtonian mechanics comes in again: Time reversing the state C', even if it is very close to A', will take the system somewhere else (say to C in the figure) which is not consistent with the description 'restored egg' and the corresponding macroscopic state will have an entropy at least as large as the oval on the right (as indicated by the dashed oval on the left). Thus, if all you know is the macroscopic state ('broken egg') then time reversal will lead to a macroscopic state with still bigger entropy than the broken egg and not back to the initial macroscopic state. This illustrates how thermodynamics can be time irreversible and still be consistent with and described by underlying time reversible Newtonian dynamics.

An example which is sometimes regarded as illustrating a different 'arrow of time" from the thermodynamic one arises in the study of electromagnetic radiation. Consider an antenna emitting electromagnetic radiation, as in radio and television broadcast. Engineers have used the very well-established theory of electromagnetism for years to compute the characteristics of the emitted waves in the course of antenna design, In that process they routinely get solutions for the waves which propagate out of the antenna, consistent with what is observed when the antenna is built and operating. But they always also get a set of solutions which are time reversed versions of the ones describing emitted waves and in which the broadcast waves are coming back into the antenna from the space in which they were dispersed in the first set of solutions. From one point of view, that is not surprising because the theory of electromagnetic radiation is time reversible, but those returning waves are (almost) never observed and are ignored by the engineers. The situation can be regarded as very similar to the thermodynamic one: To coherently arrange for a set of widely

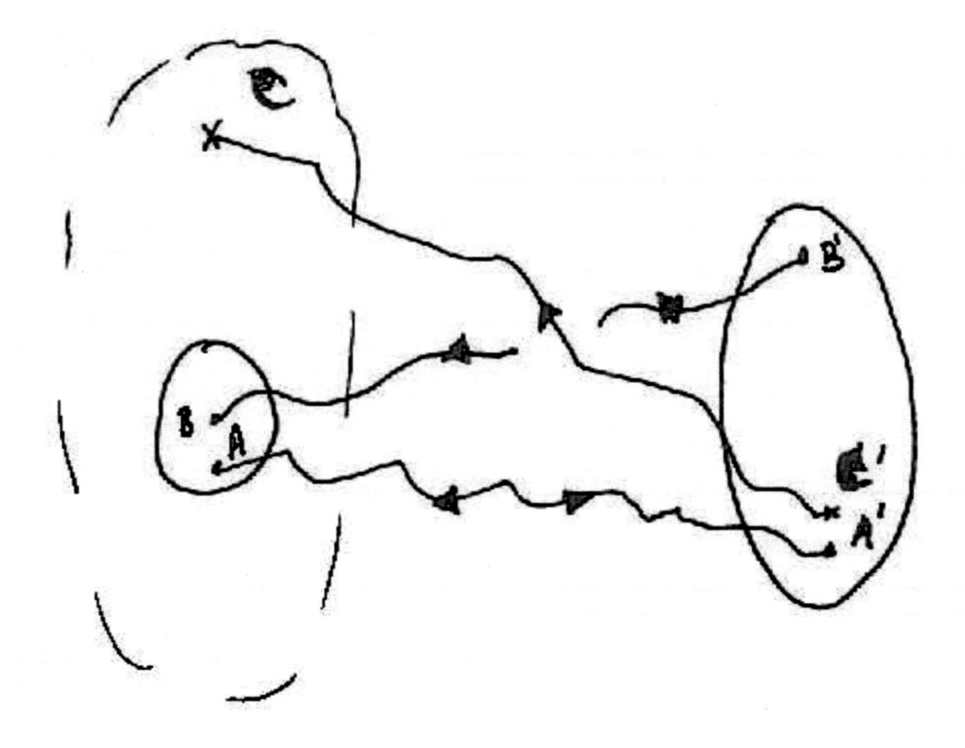

Figure 3.1 Schematic indication of how an object, such as a dropped egg, can obey time both time reversible Newtonian mechanics at the microscopic level and time irreversible thermodynamics at the macroscopic level. The small oval on the left represents the space of Newtonian states consistent with the description 'undropped egg'. A particular one of those states is labeled A. As the egg drops, the system follows the path from A to A'. The oval on the right containing A' contains all the microscopic Newtonian states consistent with the macroscopic description 'dropped egg'. It is larger than the small oval on the left because there are more possible microscopic configurations associated with a smashed egg than with an intact one. Now we imagine exactly reversing all the velocities of the atoms in the state A'. The system will return to the state A because Newtonian mechanics is reversible. However the reversal is impractical: If we try to do it, we may end up close, but not at A', say at C' and the sensitivity of the Newtonian system to initial conditions (its chaotic nature) results in the path from C' following a time reversed path back to a point C outside the oval of states describing intact eggs.

dispersed sources to coherently produce the incoming waves in the incoming solutions would be immensely difficult for engineers and extremely unlikely to occur naturally in an expanding universe in which the waves propagate quite freely into a sparsely populated environment. As in the thermodynamic case, there are occasional partial exceptions: The outgoing waves can occasionally reflect off structures such as buildings and return to the emitter.

3.4 STARTING IN A LOW ENTROPY STATE, 'IGNORANCE' AND EMERGENCE

In this discussion, I have been emphasizing two features which underlie the origin of the second law of thermodynamics: 1)The chaotic nature of the time reversible equations of Newtonian (and other) microscopic models of nature and 2) the initially low entropy state of the system. The latter point needs a little more discussion. It's ok for the egg but what about the universe as a whole? Well, we've discussed the current 'Big Bang' picture of cosmic evolution. In that picture the matter of the universe was initially (at least as far back as we can observe) at very high density. Such a state could have a very small entropy per particle, or to put it in another, equivalent, way, could be described by a much smaller number of microscopic states than the universe as we now observe it. So the current cosmological models suggest that the universe began in a relatively low entropy state and that the entropy has been increasing ever since. Most scientists thus regard the fact that the egg breaks on the floor but does not reassemble itself spontaneously as arising ultimately from the fact that the observable universe began with a Big Bang! However, details concerning how the universe evolved in the very early few minutes of the big bang and how that led to a low entropy start are not fully established. (Whether the low entropy initial state of the universe is consistent with some other aspects of the way physicists describe entropy increase mathematically in thermodynamics continues to be discussed [50].)

Some writers have contended that the dependence of the value obtained for the entropy on the coarse graining chosen makes the resulting values have a 'subjective' character and merely measure human 'ignorance'. That is misleading. The role of coarse graining in determining entropy values is closely analogous to the role of the resolving power of a microscope on the resulting image. Though a human chooses a microscope of a particular resolving power, and the image she gets depends on that choice, no one would say that the image is some kind of 'subjective' artifact. The image is an objectively real record of real properties of a phenomenon. The microscopic image could be (and often is) recorded digitally and the record will exist whether any human looks at it or not. Further, and often simultaneously, other images taken with different resolution, of the same phenomenon will describe other aspects of the phenomenon. Both images are real, and a human may, or may not, look at both or neither. Though an entropy-like calculation of the amount of information in each image can be made, neither measure will have any direct relation to the knowledge of the humans involved. Similarly, computing the entropy associated with observational data describing an object is a way of describing the specificity of the description much as the resolving power associated with a microscope image describes the specificity of the information in the image. But data recorded on the same (real) system at different length and time scales, corresponding to different coarse grainings will be associated with different entropies associated with the different specificity of description. Data sets at different time and length scales which are associated with the same phenomenon and their entropies are a measure of how much information about the phenomenon they contain, but they say nothing about any human's ignorance or knowledge.

Often the information about a physical object at one level of coarse graining reveals qualitative features which are essentially invisible, at least to humans, at less coarse-grained levels. That feature has been called 'emergence' [3, 35] and is seen in many phenomena of solid state and liquid (so called condensed matter) physics including magnetism, superconductivity and superfluidity. Those particular phenomena involve quantum mechanics but less exotic examples occur in systems in which the physics is all Newtonian, including many phenomena in fluid flow, where a kind of intermediate coarse-grained mathematical description called hydrodynamics is commonly employed to describe and predict the observed flow states and in biology. Examples of three common levels of observation and consequent description of physical systems appear in Figure 3.2. A perspective on some of the puzzling questions about the relationships of living organisms to time, such as consciousness of the present and intent for the future, can be offered from this description: Those features and others which are commonly associated with life only emerge when one observes an organism at a macroscopic scale, say of microns (10^{-6} meters) and larger. Those features lead to the biological description of the so called phenotype of the organism. It is only at that level that many features which seem to pose puzzles can be discerned. However, using only the macroscopic variables accessible at the coarse-grained level, deterministic predictions cannot be made and entropy increases with time. On the other hand, from the atomic level up, all those macrosopic features can be traced causally to the deterministic and time reversible Newtonian dynamics of the underlying atoms but deducing and predicting them from study of the motions and interactions of the atoms (mainly in polymers including proteins and nucleic acids) in the organism is extremely difficult. That is because immense numbers of atoms are acting collectively, because the mathematical apparatus used for making the predictions is extremely sensitive to small errors in the specification of the initial conditions but finally because insights from the enormous mass of data which would result from an atomic level description would be nearly impossible to obtain. We are facing the latter difficulty in recent achievements such as the successful prediction of the patterns of folding of proteins using computer codes in the discipline known as AI (artificial intelligence) which make successful predictions without revealing any general principles or insights.

The notion of emergence has only quite recently captured the attention of physicists, but the qualitative idea was understood by a school of painting called pointilism in the late nineteenth century. They made paintings using myriads of very small monocolored dots. Observed at very close range, the paintings were apparently just dots, but from a distance, scenes of trees, parks and people emerged. Successively coarser graining of the description of natural phenomena has been quantitatively useful to physicists in study of some kinds of phase transitions such as the vaporization of fluids from the liquid to the vapor state. A modern example of a pointilist rendering of an image is shown in Figure 3.3.

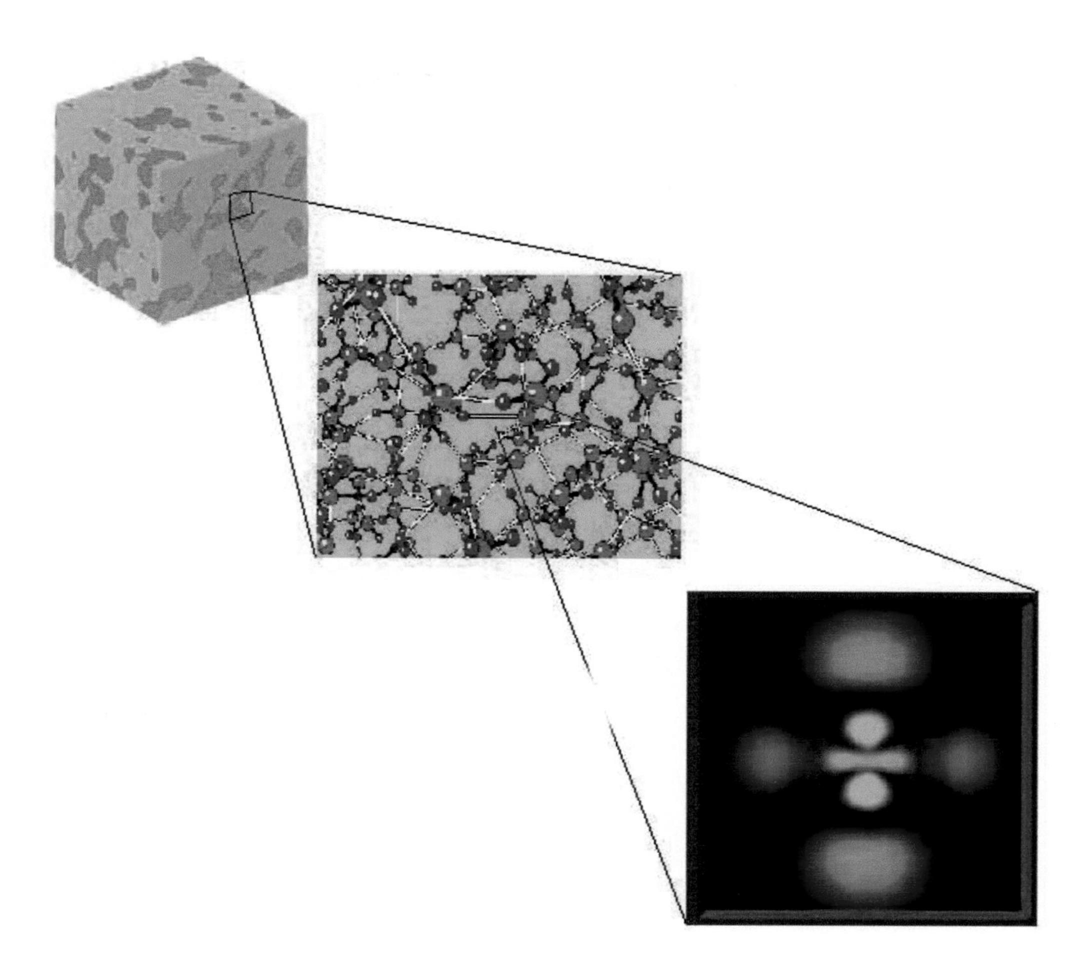

Figure 3.2 Illustration of three common coarse grainings used in the mathematical description as well as the measurement of physical objects. In the upper left the image is of a system described at a level of 10^{-6} to 10^{-4} meters, accessible to an optical microscope. The middle image (not of the same system) is an example of description at the atomic level of 10^{-8} to 10^{-7} meters, observable with xrays and neutron scattering. In the lower right is an image of the electronic structure of an atom, at scales of 10^{-10} to 10^{-8} meters accessible to recent instruments including scanning tunneling microscopes. The scales in the first two images can usually be adequately described by Newtonian physics, whereas the electronic structure has to be described by quantum mechanics, discussed in Chapter 4. One sees that qualitively, very different features emerge at the different levels. The entropy associated with the description is lowest at the electronic structure level and rises with increasingly coarse graining as one goes from the lower right to the upper left. Though the entropy is increasing, it is easier to perceive features of coarse-grained pictures such as the one at the upper left if one ignores the details of the atomic picture.

Figure 3.3 Illustration of visual emergence in the pointilist style. This is a modern rendering in which a fine grained inspection reveals only shaded dots but a coarse-grained, unfocused perusal reveals the head of the famous late 19th century statistical mechanic Ludwig Boltzmann. Boltzmann first proposed the microscopic definition of entropy which we have been discussing here. Image by Joseph Hautman, used by permission.

3.5 ENTROPY AND INFORMATION

The apparently somewhat paradoxical suggestion that humans find it easier to perceive important qualitative features of systems when they only have coarse-grained information about them is partially clarified and quantified by relating the definition of entropy to the mathematical definition of information. With that understanding, one can see how when discarding some data about a system, a clearer picture of its coarse-grained features emerges as suggested by the qualitative idea of emergence discussed in the last section.

The formal definition of information which is widely used in discussions and design of information technology at both the soft and hardware levels was first introduced by Claude Shannon, a scientist who worked in the mid to late twentieth century at the Bell Telephone Laboratories, and later at the Massachusetts Institute of Technology (MIT). Consider a 'message' which is a string of symbols. The symbols can be letters, numbers or some other set of discrete symbols. They usually stand for some physical entities, which are themselves described at a coarse-grained level. For example, in electronic computers, the symbols are often just 0 and 1, a kind of extremely simple 'alphabet', adequate for encoding any message in human language, but realized in computer hardware as describing the state, on or off, of a switch in the electrical circuitry of the computer. In the biosphere, at least two such 'alphabets' can be identified, the four nucleotides in the nucleic acids (DNA and RNA, see Appendix 1.2) and the twenty amino acids appearing in many different sequences in the linear protein molecules. And of course they can also be the set of approximately 128 symbols used in written English language, many of which represent phonetic sounds and which acquire further coarse-grained meanings related to the observed world when combined in words and sentences. Given such a set of symbols, a 'message' is just defined to be a string of such symbols. If the number of 'letters' in the 'alphabet' is denoted 'b' (2 for computers, 4 for DNA, 128 for English, etc) and the 'message' is N letters long, then the total possible number of such 'messages' is b^N. That number grows very rapidly with N (check it with a calculator) and becomes nearly unimaginably huge, for example, for a DNA molecule in a bacterium for which N is about a million.

Now imagine, with Shannon, that we receive one such 'message', containing N symbols, for example, by receiving a string of 0's and 1's from a computer in an iphone. (That is what happens repeatedly when you connect to the internet with your phone.) A preliminary definition of the 'information', denoted I, in the message is

$$information\ I = log(the\ possible\ number\ of\ messages\ of\ length\ N)$$

In some respects this makes sense: You have received only one message and you could have received any of the huge number $(b^N - 1)$ of other possible messages, so the message is distinguishing this message from a very large number of other ones and the larger the number of other ones is, the more you have learned. One can see how this works in a computer using the (so called binary) alphabet (0,1) for which b = 2. Because the definition was originally formulated with binary messages from computers in mind, it is traditional to take the logarithm in the definition using base 2. Then for a binary code the information in a message is $I = \log_2 2^N = N$, which is physically equal to the number switches required in the computer to encode the message and is known as the number of 'bits'.

However this definition does not fully capture our intuitive sense of meaningful information because of the existence of synonyms and meaningless sequences in the codes associated with any set of useful messages. In English, for example, assuming the messages are all restricted to the use of $b = 128$ symbols (only approximately right), the number in the parenthesis of a message N symbols long is 128^N but most of those sequences of symbols have no meaning in English because they do not contain the sequences of words organized in one or more meaningful sentences as recognized by the conventions of English spelling and grammar. The message has conveyed nothing to you about the fact that it is not one of those meaningless messages, because you knew in advance that it is an English message and will not contain such meaningless sequences. To convey how to correct the definition to make it more useful, I use a somewhat simpler, but less familiar example referring to the genetic code in DNA. I assume that we have a message encoded in DNA in which the 4 chemical 'letters' are describing a sequence of just 1 amino acid in a protein. In the code, as reveiwed in Appendix 1.2, which biochemistry uses to interpret such a message to make proteins, each amino acid is encoded by a 'word' which is 3 nucleotides (each one of A,T,G,C) long. So our piece of DNA must be 3 nucleotides long to code one amino acids. It turns out that evolution has ascribed an amino acid meaning to all (actually almost all, but I will simplify here) of the 64 possible sequences of 3 nucleotides, so any sequence of 3 of the four possible nucleotides can be interpreted as an amino acid. But biochemistry is only using 20 amino acids so the code described in Appendix 1.2 has a lot of synonyms. Let's suppose that our 'word' is GGG. Looking at the table (Figure 1.9A in Appendix 1.2) we see that it codes for the amino acid GLY (glycine). But there are 3 other sequences, GGC, GGT and GGA, which also code for glycine. So at the coarse-grained scale of words we have one out of $64/4 = 16$ messages, not one out of 64 so we should change the definition of information in the message to reflect that giving

$$information\ I\ in\ a\ message =$$

$$log((total\ number\ of\ possible\ messages)/(number\ of\ ways\ to\ send\ this\ message))$$

Using the properties of the logarithm this is

$$information\ I\ in\ a\ message =$$

$$log(total\ number\ of\ possible\ messages) -$$

$$log(number\ of\ ways\ to\ send\ this\ message)$$

In this example, we can think of the amino acid meaning of a 3 nucleotide 'word' as a kind of coarse-grained description of the 3 letters since there are several (4) ways to get a molecular nucleotide sequence having the same amino acid meaning. Thinking of it that way we can identify the second term in the revised definition of the information as essentially identical to the negative of the Boltzmann definition of the entropy, if we substitute the word 'system' for the word 'message'. (And use a different base for the logarithm, which only changes everything by a constant number.) Assuming for simplicity that we are using the same base of logarithms for entropy and information, we have the relation

$$S/k = log(total\ number\ of\ possible\ systems) - I$$

In DNA 3 letter example, $S/k = log(64) - log(64) + log(4) = log(4)$ as expected if we coarse grain at the 'word' level for 'word'= glycine. The general formula shows that

S/k the entropy goes down as the information goes up and when the information is maximum (only one way to send the message) the entropy is zero, consistent with what we said earlier about an atomic level Newtonian description or coarse graining a. in the dice example. On the other hand when the information is minimum it means that there are as many ways to send the message as there are messages so that I is zero and the entropy is just log(*total number of possible systems*), which is its maximum possible value (corresponding, for example, to coarse graining case d) in the example of the dice and giving $log(36)$ in that case.)

A certain amount of nonsense has been written about information and it tends to lead to some misleading impressions such as that information is some kind of a substance and has some kind of existence independent of the physical systems to which it refers. On the other hand, this way of describing entropy can clarify why humans often find that data obtained with coarser graining provides a more useful understanding of a system, as in the cases of emergent properties discussed earlier, than a more fine-grained description. As the entropy rises, the amount of information transmitted by the data goes down, making it easier for human brains to organize and comprehend.

3.6 THE SECOND LAW AND THE DIFFERENCE BETWEEN THE PAST AND THE FUTURE

Now what does this tell us, if anything, about the difference between the past and the future? On the one hand, it tells a lot because it says that the past is when the entropy is low and the future is when it is higher. But this is not very satisfactory from an intuitive point of view. Some authors [14] deal with this by separating the 'arrow of time', which is the time direction in which entropy increases from 'physical time' which, (until Chapter 5) we will consider to be Newtonian time which is measured by the various clocks discussed in Chapter 1.

Does the idea that the second law of thermodynamics provides or defines an 'arrow of time' have a useful meaning? Some authors [22] suggest that, if the universe were contracting instead of expanding then 1) the second law of thermodynamics would be reversed. That is entropy would be observed to decrease at increasing times and 2) humans would experience time backwards 'remembering the future'. There are two points here and both have been disputed. With regard to the first, it is likely that, if the contraction occurred after a long expansion, as in some cosmological models, then local effects on the rate of entropy creation would require a long 'relaxation time' before they were manifest in terrestrial phenomena. Nevertheless, reductions in entropy would be expected with increasing time after such a relaxation time had passed. However there is a question concerning whether some processes could be reversed at all, such as the formation of black holes believed to exist at the center of most galaxies. This is because, unlike systems governed by electromagnetic forces, gravitational systems are fundamentally unstable, because masses of only one sign are known.

That gravitating systems behave strangely with regard to entropy may be illustrated by a simplified version of what we understand of the history of our own universe. It is known to have begun with all the matter and radiation at high density, pressure and temperature. After ‘many adventures’ we find the universe consisting of widely separated galaxies (as well as electromagnetic radiation and conjectured ‘dark matter’ whose existence has not been confirmed). Let us consider the following cartoon-like description of such an evolution: A hot gas at high temperature T_1 at high density ρ_1 which expands at constant energy to a much lower density ρ_2. It is elementary to calculate the entropy change per molecule during such a process. It is $\ln(\rho_1/\rho_2)$ (assuming, unrealistically, for simplicity that the temperature does not change.) But the final state described by the model contains a uniform distribution of matter. It is abundantly evident, however, that what we see in the sky is not a uniform distribution of matter but a very clumpy one, consisting of widely separated galaxies (see Figure 3.4). Neglecting, in our cartoon model, the conjectured dark matter and other complications, we can model these as molecule like objects, produced by the action of gravity and consisting of N_g molecule-like ‘galaxies’ and a similar entropy per molecule estimate in that case for the entropy increase per molecule is $(1/N_g)\ln(\rho_1/\rho_2)$. The model is totally unrealistic in several ways but it illustrates that the formation of galaxies can reduce the amount of entropy increase. Note that there is a very strong dependence on the level of coarse graining chosen for the

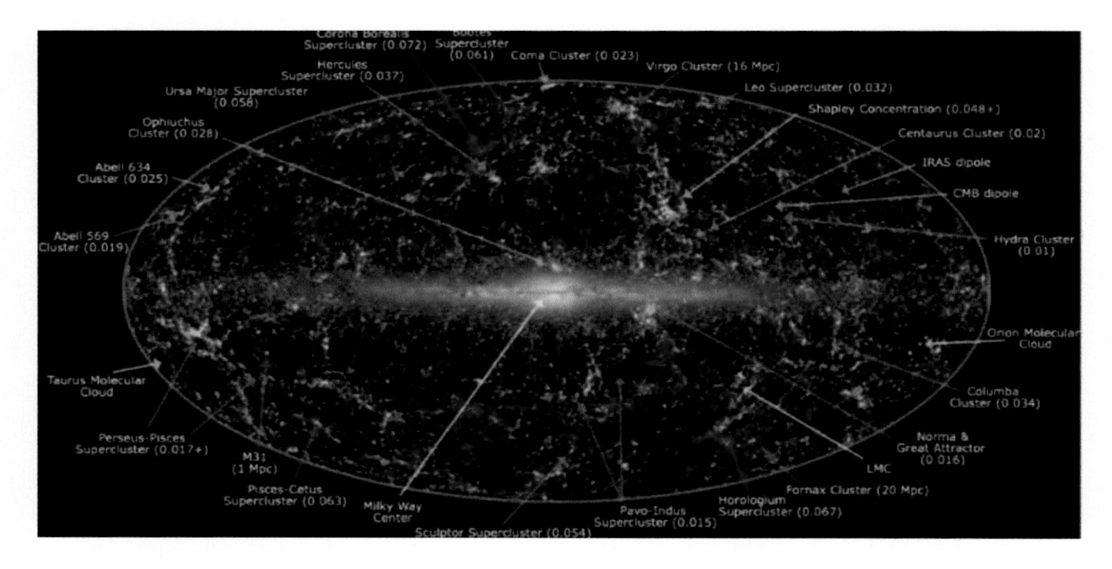

Figure 3.4 A composite image of the entire observable universe taken using a band of wavelengths in the infrared part of the electromagnetic spectrum. The spatial scale in logarithmic: Distances at the center of the picture (at the center of out own galaxy) represent much shorter real distances that distances near the edge of the picture. The picture shows dramatically that the Big Bang expansion has not led to a uniform distribution of visible matter in the universe, as the expansion of gas in a cylinder in a car or refrigerator would. That is because of the special properties of gravity, which are not significant in the case of gases in car engines and refrigerators but are significant on the larger scale of the expanding universe. From [29] by permission.

entropy calculation in such an argument: If the coarse graining scale is the same as that of the universe as a whole then one calculates that the mass density is approximately uniform and the entropy increase is $\ln(\rho_1/\rho_2)$. If the coarse graining scale is on the order of the distance between galaxies then the entropy calculation gives a diminished increase of $(1/N_g)\ln(\rho_1/\rho_2)$. If the coarse graining scale is further reduced to the average size of galaxies, then calculation must take into account that the matter has been compressed to this scale (by the gravitational field, not by externally applied pressure) leading to a DECREASE in the calculated entropy per molecule. In the last case, if the matter were behaving like gas in a cylinder with a piston, one would say that the decrease in entropy of the gas was balanced by a larger increase in the entropy of the system which pushed on the piston. In the case of gravitating systems, however, the compression occurs without an external system exerting pressure.

In more sophisticated models, cosmologists rescue the second law by assigning a large increase in entropy to the black holes which form in the galaxies during compression. The toy model described here also does not take account of the nuclear processes occurring inside stars as they form, resulting in emission of electromagnetic radiation and changes in particle number and identity through nuclear fusion which lead to entropy increases at still smaller scales. Issues concerning how to apply coarse-grained thermodynamic concepts such as entropy to systems of gravitationally interacting particles are currently very actively debated by cosmologists and theoretical physicists. The issues as understood through the late 1970's, are very clearly described, at an advanced undergraduate level, in reference [15] and in later comments by the same author [14]. The reason that these unsettled issues do not affect the applicability of thermodynamics to many phenomena is because the effect of gravity on them is very small compared to the effects of electromagnetic and nuclear forces.

Provisionally supposing that entropy would decrease during a contraction of the universe, I turn to the question of whether humans would experience the time as running in a different direction during a contraction. The only meaning that I am able to assign to the phrase 'remembering the future' is the following: In a world in which entropy is decreasing in time, the future will involve macrostates having fewer corresponding microstates than did states in the past. As a consequence, prediction of future events might be easier than it is now. Thus one would have, perhaps, a clearer view of the future. Conversely, the past, being in the same sense more complicated than the future or the present, might be harder to remember. Thus the uncertainty associated with one's psychological sense of the past and present could be reversed, with our memories being more uncertain than our predictions, rather than vice versa, so that the past might appear 'open' and the future 'fixed'. However I cannot imagine how, within existing physical law, that would lead to a psychological sense that past events occur before future events. (The issue of the order of events is revisited in Chapter 5.)

Another aspect of a conjectural reversal of the second law of thermodynamics, if it could occur, is the question of its effects on biological evolution. The evolution of complex organisms is plausibly associated with the selection pressure arising from the need to cope with unanticipated environmental shocks. If the future were to become more predictable, then the need to cope with the unexpected would be less likely to be rewarded and greater biological complexity and intelligence less favored. Thus intelligence of the sort which is usually assumed in discussion of psychological effects might be anticipated to be less likely appear at all during evolution of biospheres.

In any recurrent universe, as discussed at the end of Chapter 2, the second law would have to reverse during a part of its history: In the Newtonian description at the end of Chapter 2, the recurrent universe is in a well-defined single state throughout its history and the entropy remains zero throughout history in such a model. However in a macroscopic description, if thermodynamics as usually constituted applies, then the entropy is a 'state function' meaning that it depends only on the macroscopic state of the system and not on how it got to that state. In such a case the entropy cannot increase all the way around the loop illustrated in Figure 2.7 because it must return to the same value it had at its starting point. The entropy must decrease over part of the history. Then, in the terminology introduced above, the 'arrow of time' would be opposite in direction to the direction of 'physical time' during such periods, which might be associated with a contracting phase of the history of such a universe. (For a relativistic version of this argument see [42]; for a contrary view [15]). As in Chapter 2, I emphasize that the universe is not completely Newtonian, that even if it were, the predicted recurrence times would be many times longer than the present age of the universe, that the universe is probably not closed and that there is no empirical evidence that the expansion of our universe is slowing down. (If it were slowing down, that might lead to a recurrence.) See [22] for very interesting speculations on how one might detect evidence that the universe will follow a recurrent, closed loop path.

3.7 HUMANLY PERCEIVED ENTROPY OF THE PAST AND FUTURE

Let's consider the matter from one more point of view, namely the human perspective on time discussed earlier. Notice that the value of entropy of a system depends on the coarse graining chosen for its description. For example, in the case of a vapor, the thermodynamic macroscopic specification gave the pressure and temperature. But that choice of description was somewhat arbitrary. We could have divided up the piston into 2 regions and measured the temperature in each. Similarly we could measure the pressure in the piston at different spatial points and get a more detailed description and so on by partitioning the vapor more and more. The readings of pressure and temperature would be close to one another but not exactly the same. That kind of partitioning is used by physicists to construct theoretical descriptions known generically as 'hydrodynamics'. The difference is in perspective: We could look at the vapor with better and better 'glasses' giving better and better focus, all the way down to the molecular level. Each description would yield a different entropy because at each further focusing, the number of possible molecular states gets smaller.

Now let's consider the human perspective, discussed earlier, on the past and the future, starting with the past. From the human perspective, can we define the entropy associated with a past event? I contend we can, by analogy with the discussion above: The event is imperfectly recalled, either because memory or the records are incomplete and inaccurate. Therefore there is a large number of microscopic states associated with that past event which would be consistent with the recollections and records. We can define the entropy of the past event as the logarithm of that number of consistent microscopic states. But the accuracy and completeness of records and recollections of past events declines for events that are farther in the past. Therefore, entropy from this human perspective INCREASES as time recedes in what we usually call the negative direction into the past. What about the human perspective on the future? We have discussed the fact that human knowledge of the future (that is predictions of future states) depends on the accuracy of our knowledge of the present (velocities and positions in the Newtonian case) and is therefore uncertain because of the chaotic nature of the underlying dynamics. Let us consider a particular future state. We predict a certain configuration of matter and radiation but there is an associated uncertainty because of our inaccurate knowledge of the present state and the chaotic nature of the equations. There is a large number of microscopic future states which would give a description within this band of uncertainty. The entropy of the future human state can be defined as the logarithm of that number. But as time increases into the future this uncertainty grows and thus the entropy so defined grows with it. Human perceived entropy ALSO grows into the future. In this way, humanly perceived entropy grows both into the past and into the future. However it is not time symmetrical because the mechanisms creating the uncertainty in the forward and reverse directions are different. In the forward direction the mechanisms are similar to the ones described above for the universe as a whole, and the extent to which the growth of humanly perceived entropy of future states can be reduced is limited by the basic mechanism of entropy growth in the universe. However it can be reduced, by achieving better knowledge of conditions in the present (eg making better climate predictions by getting better, more fine grained data on the present state of the climate). In the negative direction, the mechanism is one of insufficient knowledge of the past and thus the humanly perceived entropy of the past can be reduced by study of the relics of the past (such as ice cores, tree rings and observations of distant stars whose light emissions took place billions of years ago.)

It should be noted that, in the preceding paragraph discussing humanly perceived entropy, a somewhat unusual use of the definition of entropy is employed: In the usual discussions of the second law as it applies to changes in entropy with time, it is implicitly assumed that the description of the state in question has the same level of precision (the same coarse graining) at all times. On the other hand, in the preceding paragraph I argued that in the human perception of the past and future, the level of description itself changes as one goes backward into the past, growing more coarse-grained for more negative times, and also as one looks forward into the future, where it similarly grows more coarse-grained farther into the future. Thus the growth of this humanly perceived entropy in both directions does not contradict the unidirectional growth of entropy defined at all times with the same level of coarse graining to which the second law of thermodynamics refers.

Note that, going beyond human perceptions, a similar argument relating the second law to the chaotic nature of Newtonian dynamics can also be applied to a time reversed macroscopic trajectory in which a time independent coarse graining is used:

One can reverse a Newtonian state in the present to recreate its past by changing the present sign of all the velocities and the trajectory implied by Newtonian dynamics will exactly recreate the previous history in reverse order of events. However for the same reason as in the case of the forward trajectory, a tiny error in reproducing the initial reversed state can result in a dramatically different final reversed state, requiring a macroscopic description in which the entropy is predicted to increase as the past events recur in reversed order. Thus macroscopic time reversal in the Newtonian universe would result in increased entropy as the recurring events of the past are farther and farther from the time at which the reversal occurred. In this more general argument, unlike the discussion of personally perceived entropy, the level of coarse graining was held fixed in time ,so the conclusion does not depend on a changing coarse graining in time. Furthermore, if such a reversal were achieved, empirically measured time would continue to increase in the usual forward direction, while past events occurred in reversed order. That reversed recurrence of the past would be different from the actual past to which the above discussion of human perception referred. To reverse the velocities in a Newtonian trajectory as just described is difficult and rarely done, though one can argue that it is achieved in a limited sense in so called 'spin echo' experiments [1] of various types. Such an attempt to microscopically time reverse a series of events in a small part of the universe, as in a laboratory, is an attempt to reverse time flow and will (eventually) result in increased entropy with increased time in the forward direction so the increased entropy observed in the forward direction will be consistent with the traditional statements of the second law. It is interesting that in 'spin echo' experiments that approximate a Newtonian reversal, the entropy does drop momentarily before rising in the forward time direction.

3.8 SUMMARY

In summary, the contribution of our understanding of the second law of thermodynamics to the understanding of the nature of time and in particular of the difference between the past and the future is this: The second law of thermodynamics depends on the definition of entropy, and that definition of entropy depends in turn on the level of description which is chosen for the physical system of interest. If we make a textbook (though usually implicit) choice for that level then the measured entropy of the universe is quite accurately observed to increase in time without exception. (Actually there are some caveats at small times and for systems far from equilibrium but they do not affect the basic statement.) We understand this increase in terms of the initial low entropy state of the universe and the chaotic character of the underlying dynamics which causes the future macroscopically described states to be consistent with a larger number of microscopic states than the past ones were. However if we turn to a human perspective, and use the known data about a past or future state to give the description of the state used in determining the entropy, then that humanly perceived entropy increases both into the past and into the future. This does not mean that humanly perceived thermodynamics in that sense is time reversible like Newtonian dynamics, because the mechanisms of increase in the past and future directions are different. In both directions, humanly perceived entropy may be expected to increase more slowly when (or if) human knowledge of past and present states of the universe improves. Thus both past and future will become more 'real' in a sense, or perhaps one could say that the human present will become a wider swath of time.

CHAPTER 4

Quantum Mechanics and Time

"The usual interpretation of the quantum theory is selfconsistent, but it involves an assumption that cannot be tested experimentally, that the most complete possible specification of an individual system is in terms of a wavefunction that determines only probable results of actual measurement processes. The only way of investigating the truth of this assumption is by trying to find some other interpretation of the quantum theory in terms of at present "hidden" variables, which in principle determine the precise behavior of an individual system..". David Bohm, 1951 [5]

4.1 INTRODUCTION

In the early to mid twentieth century, physicists found that they could not describe the observed behavior of the electrons orbiting the nuclei of atoms by use of Newtonian physics (also called Newtonian or classical mechanics). A new mathematical description of that behavior was developed. Called quantum mechanics, it has been tremendously successful in accounting for widely diverse phenomena at both the qualitative and quantitative level and is universally regarded as generally applicable to all matter. However quantum mechanics has some peculiar counter intuitive features and some aspects of its interpretation remain in dispute.

My object here is to clarify how quantum mechanics changes the Newtonian picture of the past, present and future and of time itself. It is important to be aware that the quantum mechanical picture reduces in most respects to the Newtonian one when the objects involved are large. Therefore the description of planetary motion, for example, is unaltered.

4.2 STATE DESCRIPTION IN QUANTUM MECHANICS

However where quantum mechanics must be used (on very small scales, with objects of very low mass or in systems at very low temperature) then the way in which physical states are described is fundamentally different in Newtonian and quantum

DOI: 10.1201/9781003037125-4

mechanics. Recall that in Newtonian physics a complete description of a state (and presumably of the universe as a whole) at a given time was specified by the positions and instantaneous velocities of all the particles in it. By contrast in thermodynamics a state was specified by just a few macroscopic quantities such as temperature and pressure. In a quantum mechanical description the state of a system at a given time is described, differently again, by a FUNCTION of the positions of all the particles and the time. Such a function is just a rule for obtaining a number associated with each possible collection of particle positions at each time. It can be described in various ways: for example, by a formula in simple cases, but also by a lookup table in a computer in more complicated ones. The positions of the particles themselves are not specified, nor are the velocities. The function, traditionally called the wave function, though it need not be particularly 'wavy', is not directly measurable but can be used to infer the PROBABILITIES that the particles in the system will have various properties, such as particular positions or velocities, in the future. Such probabilities are the best one can predict about the future in quantum mechanics. This is totally unlike Newtonian mechanics, in which the future could be definitely and precisely predicted if the present were known well enough. It is also different from thermal physics, in which there is similar uncertainty in future predictions but, because the underlying basis as we described it was Newtonian, the uncertainty could in principle be arbitrarily reduced by improving the description of the present state. In quantum mechanics there is extremely strong experimental evidence that such improvement is not possible: The wave function and the probability distributions it implies are the best one can do. I hope it is obvious that this has significant implications for ones perception of the nature of the future.

4.3 INTERPRETING PROBABILITY

If one looks a little more closely, this description poses some difficulties. Those involve the question of the meaning of probability. Probability is usually described in terms of humans making observations, of the results of a throw of the dice for example. In quantum mechanics, atomic physicists, for example, can interpret quantum mechanics in terms of human observations with no problems by interpreting the wave function as telling them the probabilities that they will observe various events in their experiments, and that works very well. But a general description of nature should not depend on the observations by a particular species of organism on a planet orbiting a particular star among more than 10^{22} in the observable universe. So in the more general context in which human observers are absent or extremely distant, what do the probabilities implied by a calculated wave function tell us about the future?

There is actually less consensus about the answer to that question than one might expect. One point of view is: Given a present state, described by a wave function at time 0, the time dependent wave function calculated for a future time describes the probabilities of the ACTUAL EXISTENCE of different configurations of the positions of the particles of the system at that later time. That may sound ok, but it is not what many physicists (probably the majority) would say. I will nevertheless pursue that point of view somewhat further before discussing alternatives. Notice that if there are, as is usually the case, no human observers available, then there is in principle some difficulty in producing an operational definition of what ACTUAL EXISTENCE might mean. But if we set that problem aside, and assume that the positions of all the particles do have such an actual existence at future times, then we have the following picture: In the quantum mechanical world there is a definite future, but it cannot be predicted. Only the probabilities of various futures can be predicted.

4.4 TWO INTERPRETATIONS OF THE MEANING OF THE WAVE FUNCTION

But let me return to the issue of what ACTUAL EXISTENCE could mean. Again, in the laboratory, there is no problem in most cases. For example, you direct a beam of electrons, which definitely behave according to quantum mechanics, at a TV-like screen. The flashes of light which you see on the screen tell you that electrons have hit the screen. If I do this experiment over and over, I will see flashes in different places on the screen even though I prepared the beam in the same way each time. Every interpretation agrees with that. An example of results from such an experiment for which the apparatus used is sketched in Figures 4.1, is shown in Figure 4.2. Each electron has encountered two 'slits' on its way from the source at the top of Figure 4.1 to the screen. If the electron were acting like a Newtonian particle it would have to go through one slit or the other on the way to the screen and, if the starting position were not well-controlled, one would see two lines on the screen after repeated experiments launching one electron after another at the screen. However, you can see in Figure 4.2 that one observes something different, namely a pattern of many lines separated by dark regions after repeated experiments (in the lower right hand corner of the picture). Quantum mechanics can be used to predict how often one will see flashes at each place on the screen (and predicts the observed pattern in the right hand corner) but it can't tell where a flash will occur in a particular run of the experiment. Everyone agrees with that too. Furthermore, by doing the experiment over and over with results shown in the lower parts of Figure 4.2 you can see a wave pattern emerging, very reminiscent of the kind of emergence we discussed as occurring from a coarse-grained description of a classical system in the last chapter. The wave pattern that emerges is interpreted to be the square of the wave function of each electron as it hits the screen.

But people part ways on the rest of the story. The interpretation discussed here now says that for each of the flashes observed a real electron followed a real, definite path from the source to the screen even though quantum mechanics cannot predict that it would take that path in advance. Many, perhaps most, physicists would not agree that the electron followed a definite path. The interpretation I am describing is sometimes called the 'pilot wave' or 'causal' interpretation of quantum mechanics. It is associated with the physicist David Bohm, but Bohm suggested several other interpretations as well. I will refer to it as the 'pilot wave' interpretation. Unlike others of Bohm's interpretations, the 'pilot wave' interpretation is mathematically equivalent to generally accepted nonrelativistic quantum mechanics and makes predictions identical to those of textbook quantum mechanics. A very good description of the pilot wave interpretation, written at an advanced undergraduate physics level,

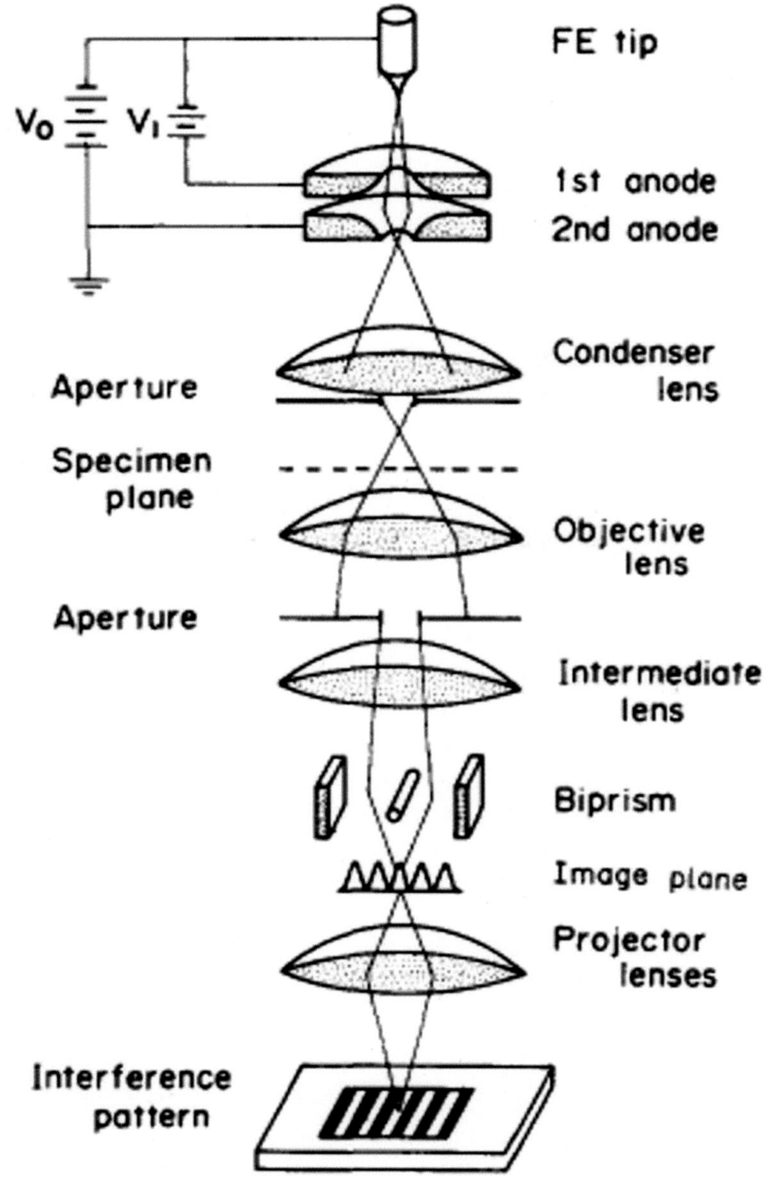

Electron-optical diagram of the interference experiment

Figure 4.1 Schematic sketch of apparatus from which the data in the next figure was obtained. Reproduced from A. Tonomura, J. Endo, T. Matsuda, T. Kawasaki, and H. Ezawa, American Journal of Physics 57, 117 (1989), with the permission of the American Association of Physics Teachers.

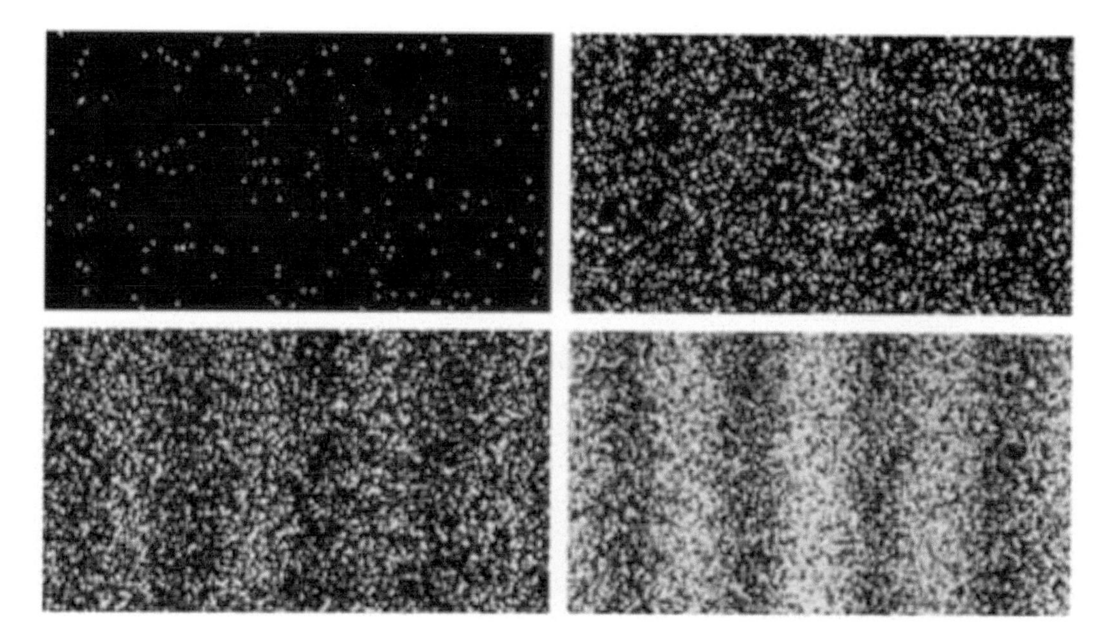

Figure 4.2 Proceeding from the upper left and from left to right, the pictures show the screen detecting electrons, each of which has passed through the double slit, after longer and longer times, when more and more electrons have been detected. Each electron is detected in a particular place, as indicated by the bright spots. The wave pattern only emerges after many electrons have been detected. It is NOT correct to interpret this as meaning that the electrons are behaving like particles at short times and like waves at long times. At all times, each electron is detected as a localized particle. The wave pattern reveals the probability, before detection, that each electron will be detected in various places. Reproduced from A. Tonomura, J. Endo, T. Matsuda, T. Kawasaki, and H. Ezawa, American Journal of Physics 57, 117 (1989), with the permission of the American Association of Physics Teachers.

is [28]. A more recent review, from a somewhat more philosophical point of view and with less mathematics is in [24]. You will not find it in standard textbooks. The latter reference is also a good place to find information about the controversies about the interpretation of quantum mechanics which continue among professionals. The reason that the formulation is called the 'pilot wave' interpretation is because the trajectories which are calculated using it, and which are regarded as the real paths followed by particles, are heavily influenced by the wave function, calculated in the standard, accepted way using quantum mechanics. The wave function is acting like a 'pilot' pushing the trajectory away from the Newtonian one it would follow in a Newtonian description of the same system. That is illustrated in Figure 4.3 where some calculated trajectories for a two slit experiment using electrons are shown, using the approach. However, the authors of [28] and [24] disagree on some fundamental aspects of the physical meaning of the trajectories shown in Figure 4.3 below. The corresponding patterns on the screen are entirely consistent with the pattern of spots

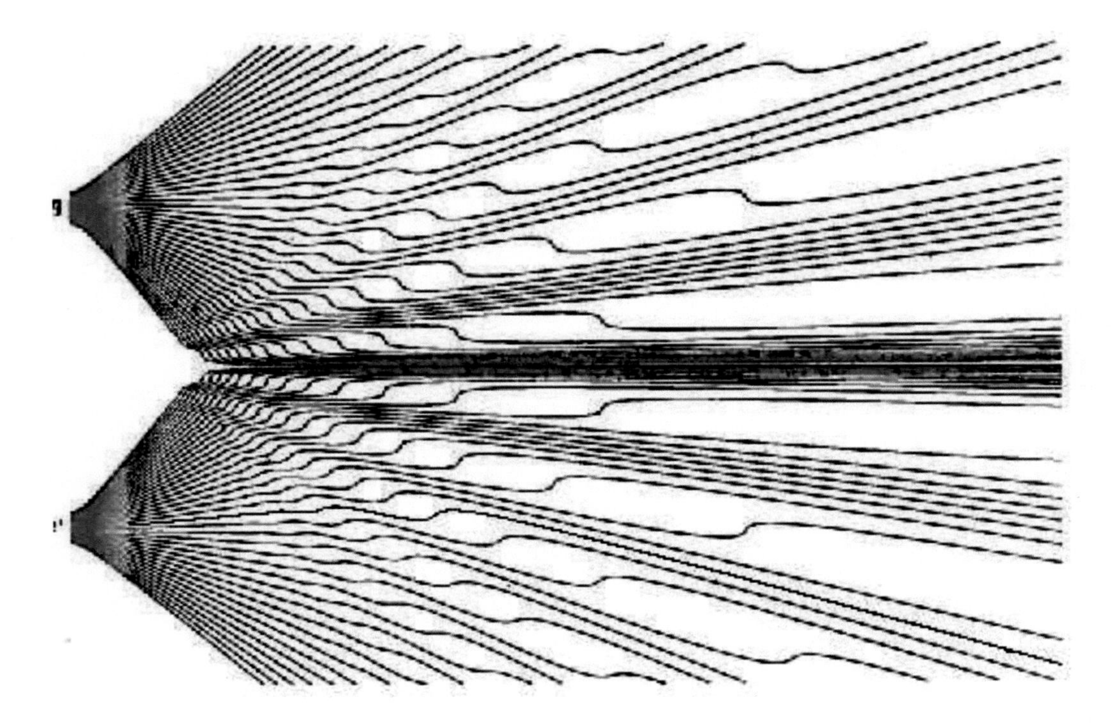

Figure 4.3 Trajectories of particles moving through two slits in a two-dimensional model of the two slit experiment, as calculated using quantum mechanics in the pilot wave approach. From [8] by permission.

predicted by the wave function calculated from standard quantum mechanics. The corresponding Newtonian trajectories would be straight horizontal lines, coming out of the two orifices and making two sharply defined spots on the screen at the right. Through the push of the wave function, the calculated electron trajectories that pass through one of the slits are strongly affected by the presence of the other slit. That is an example of so called nonlocality which is manifested in many quantum phenomena and which deeply disturbed many famous people over the last century. It does seem intuitively strange, but by accepting it, the pilot wave approach permits a realistic picture of what is going on in quantum systems to emerge. For more discussion of how the pilot wave approach deals with many famous anomalies in quantum physics, including, for example, entanglement of two particles in the wave function when the particles are far apart, see [28]. Experiments on such systems give data which agree with quantum mechanical predictions and are sometimes alleged to show that various theories in which 'hidden variables' underlie quantum phenomena cannot be true. Though one might regard the trajectories in the pilot wave interpretation as hidden, analysis shows that the pilot wave interpretation is not among the theories which are ruled out by such experiments. For one thing, the theories ruled out assume that nonlocal interactions do not occur and the pilot wave interpretation allows them. But the attributes of the theories ruled out are complex and involve other assumptions, some implicit and obscure. See [61].

Whether the pilot wave interpretation or another which I will now discuss is accepted makes no difference in the application of quantum mechanics to phenomena as observed by scientists, but it does make a difference to one's perception of time, particularly of the nature of the future. The alternative interpretation is to insist that the positions and other properties of the particles of a quantum mechanical system have no 'actual existence' at all. Only the probabilities implied by the wave function are 'real'. In the case of the electron beam discussed in the last paragraph, the second interpretation says that you cannot (should not?) attribute any path to a real electron when a flash occurs on the screen. Only the pattern of flashes has a real, objective existence. This interpretation is also consistent with the experiment since the paths are not observable no matter how clever you are at trying to see them. The mathematical description of the evolution of the wave function implies that any attempt to measure the path has effects on features of the wave function which destroy the interference pattern in the wave. Thus though you might say that the issue of which interpretation is correct ought to be easily ascertained by following the paths of the electrons as they traverse the space between source and screen, you cannot do that without changing the pattern on the screen in significant qualitative ways. Whether the paths are real or not thus remains difficult to settle experimentally. In fact recently it has become possible to learn a little about what is going on between source and detector in such experiments through so called partial or weak measurements, and further experimental input into this question can be expected. But for now, both interpretations are consistent with what is known.

But I hope it is clear that the second interpretation has a very substantial effect on ones understanding of the future. If the electron (and, by an extension which would certainly merit further discussion, all the particles in the universe) is on a definite, if unobservable path, then the future is fixed though unknowable, whereas, if, on the other hand, only the probabilities given by the wave function are fixed, then future is neither fixed nor knowable. Further, with regard to the past, the first interpretation implies a fixed past in which each particle followed a trajectory whereas the second interpretation implies no such past trajectories existed. The peculiar problems associated with making sense of the second view have led many physicists to the astonishing idea that all of the possible trajectories existed in the past and will continue into the future. This is the qualitative content of the well unknown 'many worlds' interpretation of quantum mechanics. There is no direct experimental evidence which would lead one to choose one or the other of these views, at least at the level of nonrelativistic quantum mechanics.

4.5 TIME REVERSAL AND MEASUREMENT

Other issues arise when we consider time reversal in the quantum mechanical description of nature. Recall that in the Newtonian description, the trajectories of all the particles of any physical system and their time reverses, obtained by reversing the velocities of all the particles at any moment and letting everything run backward, were equally observable trajectories in the description. Thus Newtonian mechanics was said to be time reversible. On the other hand, our experience of daily life, formalized in thermodynamic time irreversibility, is that running movies of events backward reveals sequences which do not actually occur, such as the breaking of an egg in the forward direction and its reassembly from the fragments on the floor in the backward direction. The apparent paradox was resolved, allowing Newtonian mechanics and thermodynamic irreversibility to be consistent, with the understanding that the universe is evolving from an initially very particular state which would itself be very difficult to reassemble by hand with sufficient precision to reproduce the forward time events in reversed time order. In the case of the egg, for example, the egg itself is in a very special, formally low entropy, state which is hard to reproduce, because the time reversal of its breaking is hard to reproduce by accurately directing all the pieces back into the whole egg configuration.

Now turning to quantum mechanics, there is a close similarity to Newtonian mechanics in the fact that the history of wave functions as a function of time is completely time reversible in close analogy to the history of particle trajectories in Newtonian mechanics. If a particular system's wave function can be described through forward time by a particular sequence of values, then the time reversed sequence is an equally valid and, within the theory, physically possible, sequence of wave functions. (Here I am omitting some technicalities concerning the interplay of reversal of time, change of the sign of charges and mirrorlike inversions of space which are essential for describing some phenomena involving collisions of very high energy particles.) However this does not make a definitive statement about particle trajectories and experiments observe particles, or collections of particles, not wave functions. As discussed, the wave functions are a means to predict the probabilities of such observations. The situation poses the following questions: Does something outside the theory of quantum mechanics which gives time reversible sequences of wave functions occur when observations are made and must observations be described by additional theoretical postulates? And, if so, does this make the observational content of quantum mechanics dependent on the existence of observers?

Curiously, there is not a consensus on the answers to those questions either. However most textbook accounts, as well as the early foundational work on quantum mechanics, postulated that the answer to the first question is yes: It was contended that, when an observation is made, the wave function 'collapses' to describe a state in which the position (for example) of a particle has a definite value (with probability 1) and then the time evolution of that wave function begins again. In that version of quantum mechanics, there is one part of the theory which mathematically describes the evolution of the wave function between observations and is time reversible and another part of the theory which describes measurements and for which the mathematical description is different and describes processes which are not time reversible. The textbook account, however, is widely disputed as will be discussed shortly. Turning to the second question concerning observers, the textbook description implies, though it is not usually explicitly stated, that the physical history of physical systems depends, in significant qualitative ways including time reversibility itself, on whether observations are made on it or not. This seems absurd to many physicists: While humans have a disproportionate effect on the earth's environment (in ways which have little to do with quantum mechanics) it strains credulity to imagine that they will affect so fundamental a property as time reversibility of so fundamental a theory of nature as quantum mechanics is postulated to be. Writers often refer to the panoply of issues raised here as the 'measurement problem' in quantum mechanics.

I should point out that these disputes about measurement do not significantly impede the meaningful application of quantum mechanical calculation of system properties to the interpretation of laboratory and astronomical experiments, often with astoundingly precise quantitative success. But a close look at most such situations does suggest that this is mainly because traditions have been established concerning what constitutes good technique for obtaining reproducible data from observations of systems obeying quantum mechanics. It is not because textbooks specify precisely what a meaningful measurement is or how it is to be done on a system obeying the laws of quantum mechanics. The textbooks often describe measurements in terms of mathematical objects called 'projection operators' but no laboratory contains any apparatus called a 'projection operator' and the textbooks provide almost no guidance concerning how to construct an approximation to one in a laboratory. As a result, experimental physics contains an important element of intuition and art which is much to be admired, and which has deferred the urgency of solving aspects of the 'measurement problem'.

These issues concerning measurement are not settled, but I will describe a possible way of thinking about them which I hope will make the situation seem less paradoxical. Let's think some more about the electron landing on the TV-like screen. That event is widely regarded as a typical type of measurement of a system obeying quantum mechanics and it is regarded as having measured the position of the electron when it arrived at the screen at least in the interpretation in which the electron trajectories are real but unknowable in advance. A lot is known about what happens in that event if one regards it in detail: The actual observation is the observation of a flash of light, which is recorded on the retina of the eye of a human observer or in the light sensitive material of photographic film in the old days or a photo sensitive electronic pixel in a modern camera. The electron itself usually remains in the flourescent material in a greatly altered state in which it gradually slows down due to the interaction with other electrons and the nuclei of the material. Negative charge which would build up on the screen if this process continued with many electrons hitting the screen is drained off through a circuit to which the TV-like screen is attached and which holds it at a constant voltage. In what sense are these processes time irreversible? That is, are they time irreversible because of some fundamental feature of quantum mechanics or for other, somewhat more mundane reasons? I suggest that it is the latter and that, despite the complexity of the situation, they are also completely amenable to a description of the event without recourse to an explicitly time irreversible postulate concerning the nature of the measurement process. Let us consider the emitted light as an example. As described, it ends up in a 'detector' explicitly an eye or a photosensitive material or device in a camera, where it actually disappears with its constituent energy converted to increased energy of electrons in the detector. Though it would be immensely complex and is never done, all of these processes could be followed in principle within time reversible quantum mechanics by computing the time history of an immensely complex wave function describing the coupled system of light and matter without recourse to any time irreversible, technically nonunitary, processes (I am glossing over some peculiarities of the quantum mechanical description of light here). A time reversed sequence of this time history would be a valid solution to the immensely complex quantum mechanical equation, (or actually its extension in quantum field theory to describe the light as well as the matter.) Why does the process appear to be irreversible? For much the same reason that the breaking of the egg appeared to be irreversible. It would require extreme 'fine-tuning' of the wave function describing the time reverse of the wave function describing the excited electrons in the detector at the end of the process, the light and the absorbed electron in the screen. If one could do it, one would find that the resulting system would emit light from the detector which would return to the screen where it would be absorbed by an electron which would travel back to the electron source. But one cannot do it because such fine tuning cannot be accomplished, not because a projection operator somewhat magically collapsed the wave function when the electron hit the screen.

This account suggests a description of measurement in quantum mechanics which has many of the same elements as the description of irreversibility in classical thermodynamics. It is likely, as a result, to require that in order that the measurement processes taking place in observation of systems with properties dependent on quantum mechanics be possible, the universe must have begun in a low entropy state. (The extension of the definition of entropy to quantum mechanical systems is easy to make, though I will not go into it here.) With this example, one can also try to abstract what happens in such a measurement-like process to discern what might distinguish it from other kinds of histories possible in quantum mechanical systems: There is a progression from involvement of only a few bodies such as, in the example, just one electron, to systems of great complexity and many more degrees of freedom with which it interacts. In the early days of quantum mechanics people referred to the latter system as the 'classical apparatus' and assumed that it obeyed different, nonquantum Newtonian rules. These days, physicists regard all of nature as governed by quantum mechanics and many physicists have characterized the interaction of the simpler quantum system with the more complex one as interaction with the 'environment'. Many accounts of measurement processes without the postulate of 'collapse' of wave functions contain this element of interaction with the environment. For a readable review, without much mathematics, of the long history of professional work and debate on the measurement problem from this point of view see [9].

It should be clear that, if these are the distinguishing features of a 'measurement' process, then it has nothing to do with whether humans are involved or not. This account therefore addresses the question "does the observational content of quantum mechanics depend on the existence of observers?" in the negative. Processes with the essential features of measurement can occur in any part of the universe, whether it contains any life-like beings or not, as long as the universe itself began, as evidence indicates that it did, in a low entropy state. For example, consider a cosmic ray colliding with a planet. Initially, the cosmic ray, of which there is an immense number in interstellar space, can be very well-described quantum mechanically in terms of just a few variables such as mass, energy and possibly some local electromagnetic fields.

But when it collides with the planet, a huge number of 'environmental' variables associated with the constituent material of the planet surface become involved and render the process irreversible. No humans and no biosphere are required at all. The reason that humans regard such processes as providing measurements is that a subset of them provide relatively time stable macroscopic rearrangements of matter, which can be used as a record of what happened. This is obviously what happened when the experiment leading to Figure 4.2 left the data which could be reproduced in the picture, years later. Thus we might say that measurement like processes provide humans with a means to construct a picture of the past.

To summarize this discussion of time irreversibility in quantum mechanics: The mathematical apparatus governing the time dependence of the wave functions which describe states in quantum mechanics is completely time reversible, in a way closely analogous to the time reversibility of Newtonian mechanics, though the state descriptions are different. The measurement process in quantum mechanics is sometimes described in a way (wave function collapse) which is time irreversible. However I have suggested (with others, but not all or perhaps even most physicists) that measurements of quantum mechanical systems can also be described in a quantum mechanical way which is time reversible in principle but not in practice because it involves interactions with very complex many body systems. Within such a framework, the apparent irreversibility of measurement processes in quantum mechanics has an origin similar to the origin of thermodynamic irreversibility (though different in important details) and the processes can be defined independent of any human observer. Furthermore, we can consistently regard each measurement as associated with a definite path through the immensely complicated and multidimensional space of positions of the incoming particle and all the billions of particles in the material with which it collides. That is just an extension of the 'pilot wave' description to include the particles in the many body 'detector'.

4.6 QUANTUM MECHANICS AND THE UNIVERSE

Having said this, we may ask what might be the meaning of a time dependent wave function which is associated with the universe as a whole. When we get to the large time and length scales which are involved in such a question, a realistic description must include effects of relativity but I address the question here in a model in which those considerations are postulated not to enter. There is always a huge number of variables associated with the universe as a whole and its known history suggests that in its early stages they were strongly interacting. Thus according to the concepts sketched above, no measurement like processes could take place in the early universe. It is only later , when the matter and radiation spread out in the universe, that regions of the universe, such as the particles and radiation in the interstellar medium can be described during some part of their histories as essentially independent of other parts of the universe Those parts then interact with complex systems (stars, planets) so that measurementlike processes start to take place. Does this interpretation suggest that only the wave function is real and that particle trajectories are not? Not necessarily. It appears to be logically possible that a real many particle trajectory, as well as an enormously complex wave function, are characteristic of the universe as a whole and that the trajectory is 'real' in the pilot wave sense described earlier.

With respect to multiple universes, if one considers, as I did in last paragraph, a wave function describing a (nonrelativistic) universe then in the interpretation in which various real trajectories of the particles are possible within it, the number of these possible trajectories will be immense and conceivably very diverse. In the pilot wave interpretation, the 'real' universe in which we live is moving on just one of those trajectories, which is unpredictable by quantum mechanics, but fixed and, in principle, postdictable. The other trajectories are computable but irrelevant because the initial conditions of the universe fixed on just one of them. The other interpretation of quantum mechanics, in which the wave function exists and predicts results of measurements but the trajectories do not exist, runs into immense logical difficulties in interpreting this situation. In the first place, I have indicated that by reasonable definitions, no measurement like processes are possible at all during parts of the history of the universe. So in that interpretation the wave function of the whole universe only manifests itself during those periods when it describes measurement-like events. But if we were to interpret the words 'manifests itself' as 'real' then we would end up in the absurd position of stating that the universe is only 'real' part of the time. (I'm not sure that advocates of these views interpret 'manifests itself' as 'real'.) In the pilotwave interpretation the wave function guides the trajectory of the 'real' universe throughout its history. But if one regards only the wave function itself as 'real' then one may effectively regard all possible trajectories as equally possible so that many universes simultaneously exist. Many universe advocates don't usually refer to pilot wave trajectories explicitly but to ensembles of 'alternative histories' which play a similar role in their description, though the details of their description are quite different. This is the 'multiverse' interpretation of quantum mechanics. It is quite popular among some physicists but it is hard for other physicists (including this author) to take it seriously. To say that our quantum mechanical universe is simultaneously existing in an infinitude of states which its wave function allows, seems to those who disagree with it to be a gross violation of the principle of the 'ontological' form of Occam's razor [4] which states that "Entities are not to be multiplied beyond necessity" in choosing a scientific description of nature. I note that some writers [57] have contended that the 'multiverse' interpretation of quantum mechanics is more consistent with "Occam's razor" but those authors seem to be referring to the alleged greater simplicity of the mathematics in which there is a smooth distribution of trajectories. But philosophical discussions of Occam's razor [4] do not make reference to the alleged simplicity of the mathematics of a theory but to either the "ontological simplicity" which refers to the number of real entities implied by a theory or to the "syntactic simplicity" which refers to minimizing the number of hypotheses in a theory. Multiverse interpretations fail both tests.

An analogy from Newtonian physics may clarify the point. In computing the trajectories of a planet around a star, it is very convenient to characterize the effects of the star's gravitation by a gravitational potential energy field. That field is defined so that the gravitational potential energy of any material object at a point near the star can be computed by multiplying the value of the gravitational potential energy field at that point by the mass of the object of interest at that point. From the way that the gravitational potential energy is changing near that point one determines the force on the object, such as a planet, and hence, if one knows its starting position and velocity, one can use the gravitational potential energy field to compute the planetary trajectory. But with other initial positions and velocities, the same gravitational potential energy field would lead to many other trajectories.

Now suppose a theoretical physicist told you that only the gravitational potential energy field was real, and that all the planetary trajectories possible in it exist simultaneously in the solar system, even though we only observe the planet at one particular place each time we look. I think you would say that the physicist was crazy: We can see that the planet is moving around on one trajectory, not many at once. In the (imperfect) analogy with quantum mechanics, the wave function is playing the logical role that the gravitational potential energy field is playing in the planetary trajectory case. An important difference is that in quantum mechanics we CANNOT actually see the trajectory, and hence the point of view associated with the multiverse idea is not subject to trivial refutation. But because a formulation exists in which there is such a trajectory, though unobservable, and that formulation is as good as the multiverse one at predicting observations, I contend that, on the principle of Occam's razor, the interpretation with just one real trajectory is to be preferred.

Thus I contend that quantum mechanics probably CAN be interpreted as a wholly time reversible description of nature, including measurement like processes. Quantum mechanics itself does not imply any 'arrow of time'. The future and the past can be regarded as fixed and thus, perhaps 'real', just as they are in Newtonian physics, but more is unknowable in the quantum description than in the Newtonian one. I should warn the reader that one aspect of this account is subject to a very significant criticism, arising from the fact that the pilot wave description of quantum mechanics has not, to date, been fully extended to the quantum field theory extension of quantum mechanics which is required to correctly describe phenomena involving particles moving at very high velocities with respect to one another or in very large gravitational fiels (but see [28] for a start in that direction.) Whether such an extension is possible is currently unknown, at least to my knowledge.

4.7 UNCERTAINTY PRINCIPLES

So called uncertainty principles are a much celebrated feature of quantum mechanics and are sometimes alleged to suggest that the future is not fixed. They arise from some properties of wave functions which can be proved from the mathematical apparatus which has been shown overwhelmingly to describe nature. I will first briefly discuss the first and best known example, concerning the momentum p and position x of a single particle confined to one spatial dimension. We refer to momentum instead of velocity here, and there are good technical reasons for doing so, but for the purposes of this discussion you may think of the momentum as just the mass of the particle times its velocity. The momentum is traditionally abbreviated p which might confuse a neophyte, but I conform here to convention. Now, given a wave function describing the state of this particle, it is possible, using well-defined rules, to calculate the predicted average position at which a particle will be found if one makes repeated observations of a system described by the same wave function. Notice that the average is not an average over time, for one particle, but an average over repeated experiments on many particles, one at a time, which are represented by the same wave function. That situation, for example, existed approximately for the two slit experiment and such an average would be the average over all the positions at which the electrons in succesive experiments ended up on the screen as shown in Figure 4.2. It is also possible to compute the predicted average, in the same sense, of the difference between the actual observed position and the average position, all squared. It is helpful to abbreviate, denoting the average position by $\bar{x}$ and the second quantity by $\overline{(x-\bar{x})^2}$. This quantity is the mean square deviation of the positions from their average and is often used to characterize uncertainty in all kinds of statistical data. Using the mathematical apparatus again one can go through a similar process and obtain, given a wave function, a value of the quantity $\overline{(p-\bar{p})^2}$ which is interpretable in a similar way as the uncertainty in the momentum as found in repeated observations on different particles described by the same wave function. What is remarkable is that it can be proven within the rules of quantum mechanics that there is no wave function for which both these uncertainties are zero. That can be interpreted to mean that the Newtonian program is hopeless, because you cannot calculate a Newtonian trajectory reliably without knowing both the initial momentum and position. (The pilot wave trajectories described earlier are not Newtonian.) That uncertainty relation is entirely consistent with all experimental data, but it does not say anything about the future because we are not attempting within the quantum rules to compute what happens in the future from Newtonian dynamics anyway. The reason that the uncertainty does not prevent us from calculating useful Newtonian trajectories for massive objects like billiard balls is because the uncertainties can be made extremely small for such objects. Therefore the uncertainty relation relating momentum and position does not have any direct qualitative effects on what has been said already about the nature of the future in a quantum mechanical world.

On the other hand Heisenberg, who was one of the inventors of quantum mechanics and the discoverer of the position-momentum uncertainty relation, also stated that there is a similar constraint on times when events take place and the energy of a particle. The mathematical status of this alleged principle is less clear: Time does not play the same mathematical role in quantum mechanics that position and momentum do and the proof of an uncertainty relation between time and energy, cannot be simply achieved by imitation of the procedures used to prove the position-momentum relation. However there are features of quantum mechanics and related physical phenomena which do suggest such a relation and a general result was derived in 1945 by the Russia scientists Mandelstam and Tamm [32]. The time involved in the relation is not the time appearing in the argument of the wave function, but is the time which a particular, time dependent, process takes on average. A well-known example is the decay of radioactive nuclei, which we have discussed in Chapter 1 when describing methods for measuring the times when events in the deep past occurred. That process was characterized by an average decay time which we called the half life. The principle proved showed that the uncertainty, over many measurements on different nuclei of the same type as defined above, of the energy of the resulting nucleus plus other decay products could not be smaller than (a constant)/(the half life). What it means qualitatively is that the longer the half life is, the better successive measurements of the energy of the decay products of different nuclei of the same type will agree with one another. This does not make a statement about the future of a particular nucleus but it does make a statement about the future properties of an ensemble of them. It puts limits on what we can predict about the future, because such accessible predictions are only available from information in the wave function, independent of any interpretation concerning an underlying trajectory or its absence. Thus it illustrates the general idea that the future can be definite and fixed but unpredictable. Note that the time involved in the energy-time uncertainty relation depends on what dynamical process is being described by the wave function, whereas the position-momentum relation is independent of what kind of process is being described. The precise general specification of the probability distribution of the times of occurrence of predicted events in quantum mechanics has been a subject of ongoing professional concern and debate [38]. See [28] for a proposal using the pilot wave formulation of quantum mechanics.

CHAPTER 5

Relativity and Time

"Time," he said, "is what keeps everything from happening at once." Ray Cummings, "The Girl in the Golden Atom", All-Story Weekly (1919) sometimes attributed to Albert Einstein

"..the supreme goal of all theory is to make the irreducible basic elements as simple and as few as possible without having to surrender the adequate representation of a single datum of experience." Albert Einstein, "On the Method of Theoretical Physics," the Herbert Spencer Lecture, Oxford, June 10, 1933, often paraphrased as "Things should be as simple as possible, but not simpler"

5.1 INTRODUCTION

A few decades before quantum mechanics was found as a way to describe non-Newtonian behavior of particles on very small scales, physicists found another problem with the Newtonian description of particles moving at very high velocities. The solution to the problem was the special theory of relativity, which describes how the space and time separations of events differ when recorded in frames of reference moving at rapid velocities with respect to one another. There is a second aspect to relativity, called general, as opposed to special, relativity, which is basically a theory of gravity, also differing from Newton's description but reducing to it in low gravitational fields and velocities. In this chapter I will be mainly concerned with the special theory.

5.2 WARPING NEWTONIAN TIME

To try to make these matters as clear as possible, I begin with the description of a wave moving on water, say on a lake, with some velocity, which I will call c_{water}, in a straight line (no curves). Now we imagine that the wave is passing by a dock and that a person has set up a measurement scheme to measure the velocity of waves moving past the dock. (See Figure 5.1.) The person has made two marks on the dock, a distance, call it Δx, apart. She also has some kind of stopwatch in order to measure the time, call it Δt, between the moment when a wave passes the first mark and the

DOI: 10.1201/9781003037125-5

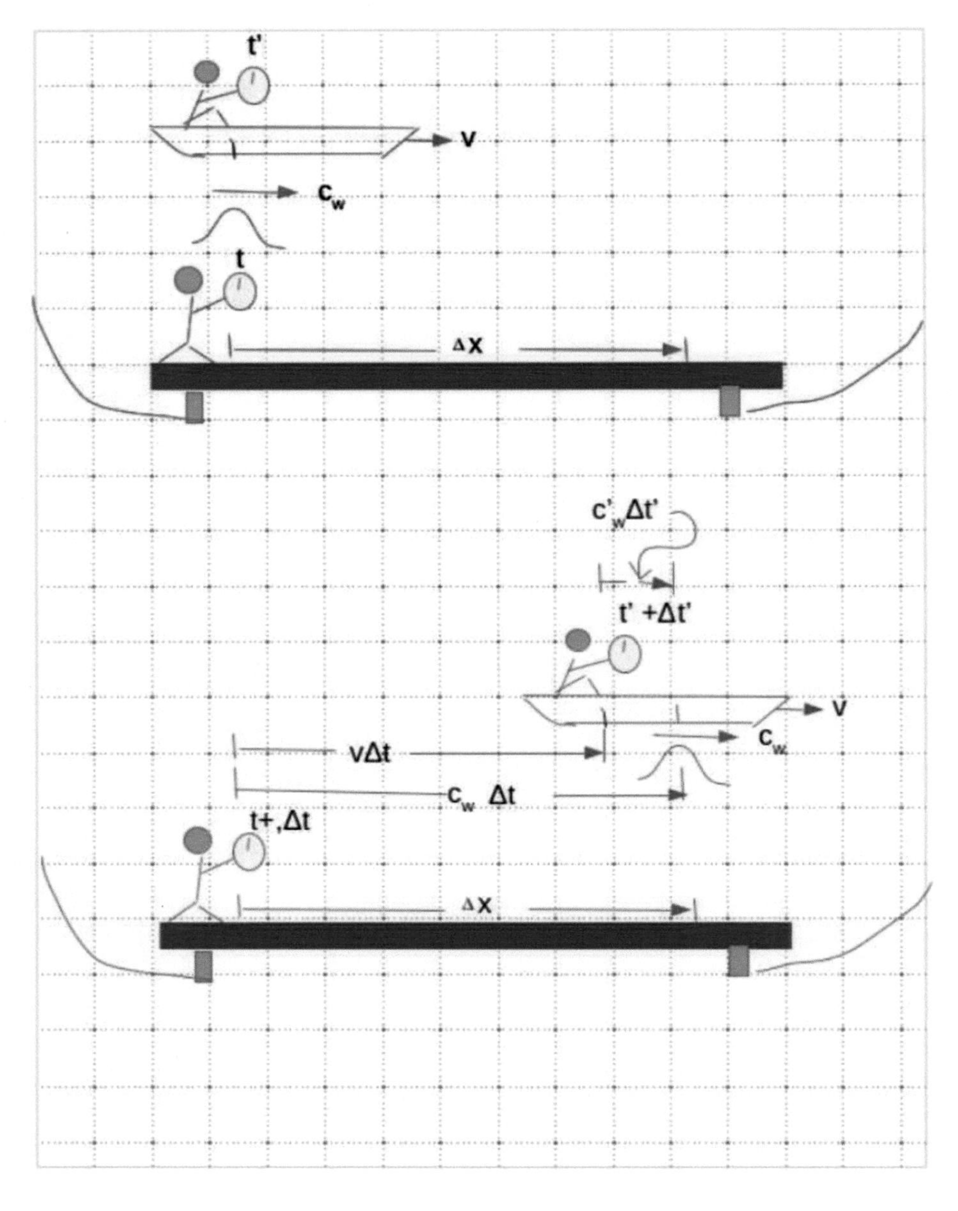

Figure 5.1 Illustration of measurement of the velocity of a water wave. The time and space separations between two events, namely the arrival of the peak of a water wave at two points marked along a dock (brown). The top panel shows the situation when the wave peak arrives at the first mark on the dock and the bottom panel shows the situation when it arrives opposite the second mark on the dock. The wave is assumed to have speed c_w and the canoe is assumed to have speed v. By making marks on his canoe when the wave passes the first mark on the dock and a second mark on his canoe when the wave passes the second mark on the dock, the canoeist deduces that the wave has moved a distance $\Delta x'$ with respect to his canoe which is less than the distance Δx which it has moved along the dock. In the Newtonian picture, the two time separations Δt and $\Delta t'$ between the two events as measured by the person on the dock and by the person in the canoe respectively are assumed to be the same.

time when it passes the second mark a distance Δx away from the first mark along the dock. She simply starts the stopwatch when the wave passes the first mark and stops it when the wave passes the second mark. The person on the dock then has enough information to calculate the speed of the wave ($c_{water} = \Delta x/\Delta t$).

Now suppose that it happens that at the moment that the wave passes the first mark on the dock, a boat, say a canoe, happens to be passing along the dock so that the wave is between the dock and the boat and the wave are moving in the same direction. A person on the boat also has a stopwatch and a magic marker. When the wave passes the first mark on the dock, he makes a mark on the side of the canoe indicating the position of the wave at that moment and starts the stopwatch. When the wave gets to the second mark on the dock he stops his stopwatch and makes another mark on the boat opposite the wave at that later moment. Then he records the time $\Delta t'$ on his stopwatch and measures the distance $\Delta x'$ between the two marks on his boat. He gets a number for the velocity of the wave by calculating $c'_{water} = \Delta x'/\Delta t'$ using his measured numbers. It is important to understand here that given the Newtonian assumptions about time prevalent for 300 years in physics as well as in most peoples' intuitions to this day, the two numbers read from the stopwatches Δt and $\Delta t'$ should be the same, because they measure the time between the same pair of events ('wave passes first mark on dock' and 'wave passes second mark on dock'). But if it is not already clear to you, you can see by studying the diagram that Δx is not expected to be the same as $\Delta x'$. In the case shown in the diagram, $\Delta x'$ is expected to be smaller than Δx because the boat has moved along some distance, following the wave, during time Δt. Correspondingly, the wave velocity calculated by the person in the boat from his measurements will not be the same as the one obtained by the person on the dock, and will be smaller. If the boat is moving at speed v we have the relations

$$\Delta t' = \Delta t$$

and

$$\Delta x' = \Delta x - v\Delta t$$

and the velocities of the wave as measured by the two observers are

$$c_{water} = \Delta x/\Delta t$$

and

$$c'_{water} = \Delta x'/\Delta t'$$

putting the expressions for $\Delta x'$ and $\Delta t'$ above back into these expressions for the velocities you can see quite easily or with just a little algebra that we expect

$$c'_{water} = c_{water} - v$$

that is, the wave as observed from the boat is perceived to be moving slower than the wave as perceived from the dock. Indeed if the boat were riding the wave, then the wave would not be observed to move along the boat at all and the measured c'_{water} would be zero, consistent with the last equation (put in $v = c_{water}$) so this seems to make good sense.

These relations between times and velocities as measured from one point of view and from another in which the observation is made from a platform moving with respect to the first with speed v are called Galilean transformations. That is because Galileo wrote them down in his preNewtonian studies of motion (which were consistent with Newton's, though not as complete.) Almost everyone in the late nineteenth century (more than two hundred years after Newton) assumed that they would also be true for light waves moving through empty space as well as for every other kind of wave such as water waves moving on the surface of water.

So physicists decided to use these relations to determine how fast the earth was moving through the medium of empty space (which they called the 'ether') as illustrated in Figure 5.2. In appendix 5.1 I describe some details of the most famous such experiment, which was first reported by physicists Michelson and Morley working in the United States. The result of the experiment, which was repeated many times by different observers, was that the speed of light was totally unaffected by the velocity of the earth around the sun. In the language of the example of the boat and the dock they found that $c'_{light} = c_{light}$ where the unprimed velocity c refers to the speed of light as measured from an imagined platform which is stationary with respect to the 'ether' and the primed c is the velocity of light as measured on the earth moving at speed v through it.

The result astounded and perplexed physicists of the era. Michelson wrote to a friend that his experiment had 'failed'. Many explanations were offered but what was perhaps the most radical one, proposed by Einstein, turned out to be the one which is accepted in our era and has had millions of experimental confirmations. It challenges a basic assumption about time which we have been assuming so far in this book. In terms of the Michelson Morley experiment, what Einstein said was that if the two light speeds c'_{light} and c_{light} as measured from the different platforms were the same then since $c'_{light} = \Delta x'/\Delta t'$ and $c_{light} = \Delta x/\Delta t$ and we already know that $\Delta x'$ can't be the same as Δx, then times $\Delta t'$ and Δt must also be different so that the ratios, which are the measured velocities, come out the same. It means that measured time intervals between two events must be different when measured from different platforms if they are moving with respect to one another. (In what is called special relativity, which I will mainly describe here, the velocity of one platform, or 'frame' with respect to the other is assumed to be constant. For many experiments this is a good approximation. Einstein's general theory deals with the case in which the 'frames' can be accelerating with respect to one another.) To use the general notion that the time intervals could be different to get a result consistent with the experiment on light velocities Einstein rewrote the Galilean relations as follows

$$\Delta t' = G(\Delta t - B\Delta x)$$

and

$$\Delta x' = G(\Delta x - v\Delta t)$$

which are to be compared with the Galilean relations

$$\Delta t' = \Delta t$$

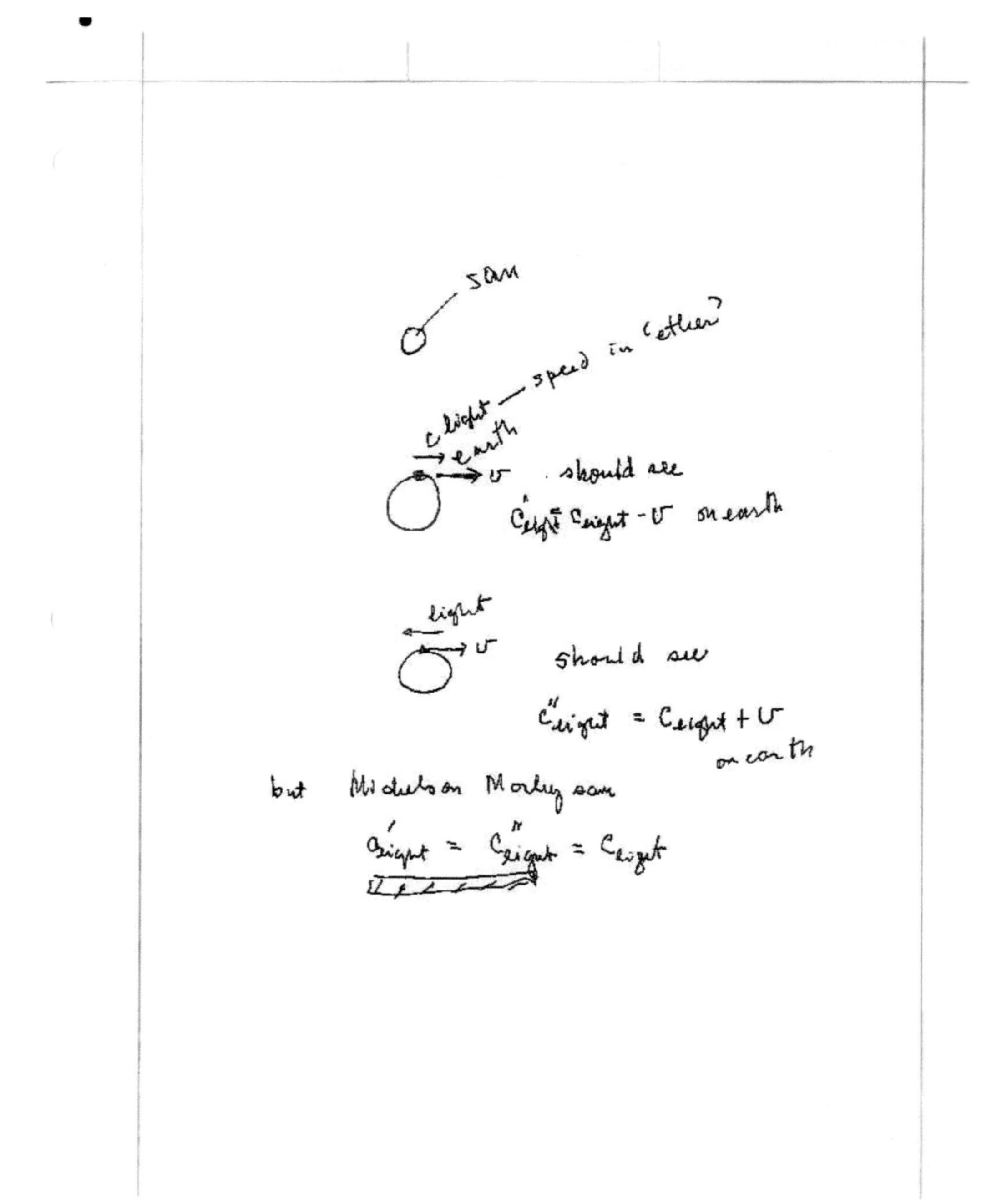

Figure 5.2 Illustration of the idea of the Michelson and Morley experiment on the velocity of light.

and

$$\Delta x' = (\Delta x - v\Delta t)$$

There are two changes. The second equation, for $\Delta x'$, is the same except for the so far unexplained factor G in front. The first equation, for $\Delta t'$ has changed more. It also has a factor G in front but it is also assumed that the distance Δx between the two events in the first frame can affect the time $\Delta t'$ as measured between the two events in the second frame and the new term has a second unexplained factor B in front of it. (If you read a standard textbook on this subject you will usually find that the Greek letter γ is used in place of G. I don't want you to be distracted by too many Greek letters.)

We first explain how the factor B can be chosen to make these relations consistent with the result of the Michelson Morley experiment, or as often described in textbooks, consistent with the assumption that the velocity of light is the same when measured in any frame. For that purpose, we suppose that the two events in question are the arrival of the same light pulse first at one point in space and then at another a distance Δx away as measured in one frame and $\Delta x'$ away as measured in the other frame (just as in the water wave experiment). The velocity of the light pulse should then be, if the two measured wave velocities are the same,

$$c = \Delta x/\Delta t = \Delta x'/\Delta t'$$

giving, with Einstein's assumed forms:

$$c = \frac{G(\Delta x - v\Delta t)}{G(\Delta t - B\Delta x)}$$

Canceling out the G's and using $c = \Delta x/\Delta t$ on the right gives

$$c = \frac{c - v}{1 - Bc}$$

which is easily solved for B with the result $B = v/c^2$.

This assures that the relations give the Michelson Morley result for light and you can work out that, if $v/c << 1$ as for a water wave, it will also give the previously described Galilean result for the velocity of a low velocity wave like a water wave to an excellent approximation..

So why do the relations have the extra factor G in front (it cancelled out in the velocity calculations) and what is it? The answer reveals other motivations and requirements placed on the new relations between space and time: First, Einstein wanted the relations to be true for the time and space separation of *any* pair of events and not just the arrivals of light pulses at various places. The argument is that if time measurements are different between events as measured in different frames, it shouldn't matter on which pair of events you are making the measurements. That has two consequences, In the first place it would have to mean that the c in the equations is not fundamentally arising from some special property of light, but is a property of space and time itself. In that interpretation, light has that speed only because light waves have no mass and not because of any other special properties of light.

Secondly, Einstein imposed one more requirement and that is why the theory is called a theory of 'relativity'. He required that the same equations work whether they refer to measurements of the pair of events as measured in the primed frame in terms of the measurements made in the unprimed frame or, on the other hand to measurements made in the unprimed frame in terms of measurements made in the primed frame. The only difference should be in the sign of v: If it is positive in the first case it will be negative in the second case. Think of the boat, the dock and the water wave. If we are on the dock, the boat 'frame' is moving to the right with speed v. If we are on the boat, the dock is moving backward, that is with velocity $-v$ with respect to us. These considerations lead to two sets of equations, which have to be equivalent if the theory is to work in the same way when deducing time and space intervals in the primed frame from the unprimed one or deducing time and space intervals in the unprimed frame from the primed one.

$$\Delta t' = G(\Delta t - (v/c^2)\Delta x)$$

$$\Delta x' = G(\Delta x - v\Delta t)$$

and

$$\Delta t = G(\Delta t' + (v/c^2)\Delta x')$$

$$\Delta x = G(\Delta x' + v\Delta t')$$

It turns out that there is only one value of the square of the so far mysterious factor G which will make these two sets of equations consistent. To find that value algebraically is a little more complicated than the algebra leading to the expression for the factor B and I have put it in Appendix 5.2. The result is

$$G^2 = 1/(1 - (v/c)^2)$$

Looking back at the equations describing the transformations, we see that G, not G^2 enters the transformations so we must take the square root. Two other issues then arise. 1) If v were greater than c then the square root would be a physically meaningless imaginary number. Thus the requirements of relativity and the fixed velocity of light in all frames lead to the surprising, though now familiar, conclusion that frames cannot move with respect to one another at more rapid uniform speeds than c. That is abundantly confirmed in experiments. 2) Taking the square root when $v < c$ we obtain

$$G = \frac{\pm 1}{\sqrt{1 - v^2/c^2}}$$

So we have two possible values of G. Which should we choose? The answer is determined by insisting that we get back to the Galilean transformations discussed in connection with the example of boat, dock and water wave when v is much smaller than c (but not zero). The final assumption leads to the conclusion that only the plus sign will work and we have a unique value of G:

$$G = \frac{1}{\sqrt{1 - v^2/c^2}}$$

Putting it back in the transformations we get the famous Lorentz transformations of Einstein's theory of special relativity

$$\Delta t' = \frac{(\Delta t - (v/c^2)\Delta x)}{\sqrt{1 - v^2/c^2}}$$

$$\Delta x' = \frac{(\Delta t - v\Delta t)}{\sqrt{1 - v^2/c^2}}$$

(They are called Lorentz, not Einstein, relations because another physicist wrote them down first, while not giving them Einstein's, then radical, interpretation.)

Summary: To deal with the problem of the observed constancy of the speed of light in all frames we followed Einstein's assumption that the way of transforming space and time intervals measured in different frames moving with a constant velocity v with respect to one another could not follow the Galilean transformations which had worked well for many phenomena when v was small, but must be altered in the way presented, in which the time separation of two events in one frame of reference depended on both the time separation and the space separation in the other frame of reference. Requiring that the speed of light c come out the same in both frames, as observed, then permitted one of the constants (B) in the proposed new transformations to be determined. We then added the requirement that the transformations work both forward and backward, that is, that one could as well assume that the frame originally assumed to be moving with v was stationary and that the one originally assumed to be stationary was moving at velocity $-v$ and one should get the same result. That led to a definite expression for the other constant (G) in the proposed transformations as long as we required that we recover the Galilean transformations when $v << c$.

I hope you can see that, though some algebra was required, the assumptions made are quite simple, transparent and not very numerous. The most surprising and nonintuitive assumption is that the speed of light is the same in all frames of reference. That assumption was forced upon the physics community by the results of repeated experiments. The relativistic requirement, sometimes stated as requiring that no frame of reference is preferred or special, is intuitively quite natural and is also obeyed by the Galilean transformations. However the final consequences for one's conception of time are quite profound. I will next discuss some examples of the effects, focusing on experiments which especially bear on our conceptions of the nature of time.

Before that, I add a few asides intended to help those reading elsewhere about relativity to relate this account to others: Though this account is elementary and can be followed by anyone with a little knowledge of high school algebra, it is absent from most textbooks, which often simply state the assumptions and the resulting Lorentz transformations without indicating much, or anything, about how one gets from one to the other. Also, I have not made much mention of 'observers' here (though the individuals on the boat and the dock could be described that way), because the theory does not refer in general to what humans do while making measurements but to how the counters, such as oscillating states of cesium atoms, decaying particles, or the rotations of a planet, which we use as clocks, behave differently in different reference frames . Further there was no reference to rockets or space travel. Many elementary accounts of relativity refer to rockets and space travel but many of the associated 'thought experiments' have not been done and are in many cases unlikely to ever be done. Such accounts can leave a somewhat fantastical impression on an unprofessional reader. In the description of experiments to follow I have chosen examples which have been done or could quite easily be done with present technology. Some points in the arguments given here, particularly about the assumptions leading to the choice of which root of G^2 to use, appear very rarely in textbooks, though I did find them in the superb book by the eminent late physicist Wolfgang Pauli [46]. Finally I have not discussed the implications of these results for the energy and momenta of particles. Those implications play a huge role in interpretation of many experiments but they bear less directly on the challenges which the theory poses to our conception of time.

5.3 TIME DILATION

The effect of the Lorentz transformations on the behavior of clocks is manifested in dramatic fashion every day in experiments carried out at particle accelerators throughout the world. However, I will here describe a case in which the acceleration to high velocities occurs naturally without human intervention, and in which it is quite easy to see that the relativistic effects occur whether there are human observers, humanly contrived clocks or other human artifacts present or not. The experiment has been reported in detail at least twice [49], the second time [21] for mainly pedagogical purposes and is repeated in some form by undergraduates majoring in physics in many universities. It concerns the radioactive decay of a short lived particle called the muon or formerly, the mu-meson. The decay of the muon to other particles is much like the decay of radioactive nuclei which was discussed in Chapter 1 where I described how such decaying particles could be used as clocks to measure the age of ancient rocks. Those radioactive decays were described in terms of a half-life which specified the time required for half of a sample of the particles in question to decay. The muon behaves in that respect in the same way, but it has an extremely short half-life, compared to the isotopes discussed in Chapter 1, of 1.56×10^{-6} seconds. Despite the difference in time scales, we can use the decay of a batch of decaying muons as a clock, much as the isotopes discussed in Chapter 1 are used, but to measure shorter times. Most of the muon's properties are irrelevant to this discussion, but to avoid mystification I will mention that it has many properties like those of the electron, for example the same charge and spin, but a much larger mass, and it decays very quickly after its production (with the energy dumped into electromagnetic radiation and neutrinos) while the electron is stable and never decays. For the purposes of this discussion we may accurately think of the muon as a radioactive particle decaying with the same decay law which characterized the radioactive nuclei used in determining events occurring in the early earth as discussed in Chapter 1. That law could be expressed in terms of the half-life τ of the radioactive particle in question as

$$N(t) = N(0)(1/2)^{t/\tau}$$

Here $N(t)$ is the number of surviving particles in a sample at time t and $N(0)$ is the number of particles which were present in the sample at time $t = 0$. You can see using the formula that when t is equal to one half-life τ that it predicts that half the original particles will remain, that after two half-lives $1/4$ will remain, and so forth. Muons behave in exactly the same way except that instead of half-life of thousands to millions of years which characterized the nuclei described as useful for measurements of deep time in Chapter 1, in the muon we have a particle with a very short half-life of 1.52×10^{-6} seconds as measured already in many laboratories on earth by the time the experiment I will describe was first done. We will think of a cloud of muons created at a particular time as acting like a clock, much as samples of radioactive isotopes in geological samples have been used as clocks to determine the ages of the samples. The experiment to be described utilized the fact that, as a result of the influx of other, longer lived, high energy, particles on the upper atmosphere of the earth, muons are continually created and rain continually down toward the surface of the earth. It was also known that the creation process gave those muons extremely high velocities of about 99% of the speed of light as they came down. In the experiment described in reference [49] the experimentalists took detectors of muons to two sites in Colorado, one at a place called Echo Lake which was high in the Rocky Mountains near Denver and another in Denver. The detector at Echo Lake was 1624 meters in altitude above the one in Denver. Knowing the speed with which the muons were coming down, they could estimate that a muon passing vertically downward through the elevation of Echo Lake would take

$$1624\ meters/v = (1624\ meters)/(3 \times 10^{8} m/s \times .99) = 5.47 \times 10^{-6}\ seconds$$

to get from that elevation to Denver. They measured the number of muons found per unit time and per unit area in both detectors and took the ratio of the results getting

$$N(t)/N(0) = 0.88$$

where $N(t)$ was the number detected in Denver and $N(0)$ was the number detected at Echo Lake. $N(t)$ was smaller than $N(0)$ because the muons had more time to decay before they reached Denver.

Now we can estimate what that ratio should have been if we neglect the effects of the Lorentz transformation on the time by using the equation defining the half-life as

$$N(t)/N(0) = (1/2)^{(5.47x10^{-6} seconds/1.52 \times 10^{-6} seconds)} = (1/2)^{(5.47/1.52)} = .08$$

which is about 10 times smaller than what was observed. This illustrates that neglecting relativistic effects will not work with a particle moving at such a high velocity. To apply the Lorentz transformation to this experiment and include the special relativity effects we must note that, in the laboratory experiments which determined the half life of the muons, the muons were almost at rest with respect to the laboratory (on earth) rather than tearing down toward the detectors at 99% of the speed of light. In the case of the muons which are tearing down toward the detector in Denver, we consider an unprimed frame of reference attached to the moving muon. In that frame attached to the muon the half life is the same as the one measured in the lab and we denote the time for the muon to get from Echo Lake to Denver as it would be determined in that frame as Δt. We denote the frame of reference attached to the detector in Denver as the primed one. The calculation of the time for the muon to get from Echo Lake given above was made from the earth frame and, denoting it $\Delta t'$ we have $\Delta t' = 5.47 \times 10^{-6} seconds$.

$$\Delta t' = \frac{(\Delta t - (v/c^2)\Delta x)}{\sqrt{1 - v^2/c^2}}$$

What's Δx? It is the distance that the muon moved in the unprimed frame while traveling from the elevation of Echo Lake to the Denver detector. But, by definition, the muon didn't move at all in that frame so $\Delta x = 0$. Thus the time which we should use in the equation defining the half-life in this experiment is Δt and it is related to the time $\Delta t'$ as measured from the earth by

$$\Delta t = \Delta t' \times \sqrt{1 - v^2/c^2} = 5.47x10^{-6} \times \sqrt{1 - .99^2} = .77 \times 10^{-6} seconds.$$

Much less time has passed during the trip down in the frame of the muon. The fraction of muons remaining at Denver should therefore be

$$N(t)/N(0) = (1/2)^{\frac{.77 \times 10^{-6} sec}{1.52 \times 10^{-6} sec}} = .70$$

which is 8.75 times larger than what was given by the nonrelativistic calculation and was judged to be consistent, within the uncertainties in the experimental observations, with the observational result ot 0.88. (I have simplified the discussion in the reference by Rossi and Hall and have used the presently accepted value of the half-life of the muon in the discussion. The half-life value used by Rossi and Hall was slightly different. In the paper by Rossi and Hall and the similar experiment by Frisch and Smith it was necessary to be careful to count only mesons with the stated velocity and that required significant experimental complications. Versions of this experiment are done by physics students at advanced undergraduate level in many universities.)

The take home message here is that the Lorentz transformations imply that the 'clocks' in rapidly moving frames (such as that of the muon where the muon itself is acting as a clock) run slower than clocks in a relatively stationary frame (such as the laboratory in the muon experiment) where stationary muons decay faster. It is a good example because it illustrates that the effect has nothing to do with the construction of clocks or with humans reading them or not. It is a statement about the rates at which the regular repeating events which we use as clocks themselves act differently in different frames. The same number of muons would get to Denver whether there was a detector there to count them or a human to read the detector or not. It is very odd but, again, it has been confirmed in millions of experiments.

5.4 SIMULTANEITY

Now I consider an imaginary observation, which however could be done quite easily with presently available technology. Suppose that an astronomical observatory on

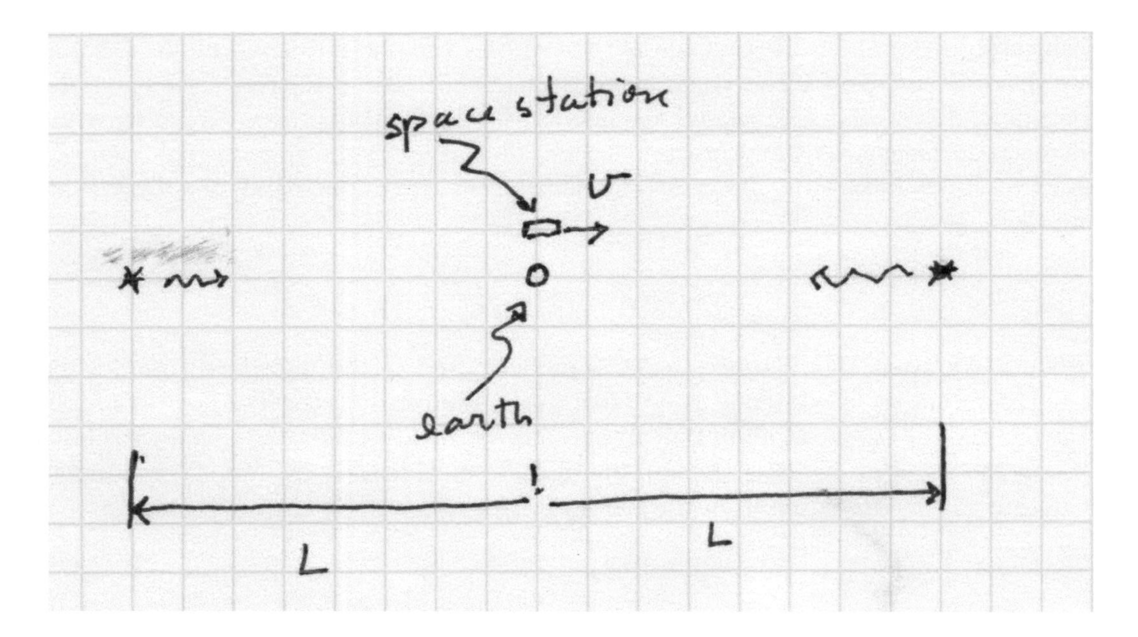

Figure 5.3 Imagined observation of two supernovae from different frames of reference.

earth detects pulses of light from supernovae in two galaxies which are the same distance, call it L, from the earth but in opposite directions (see Figure 5.3).

The pulses arrived at the same time and the distances were the same so one can infer from those detections that the pulses were emitted at the same time. In the frame of the earth, the separation in time between the two pulse emissions was $\Delta t = 0$ and separation in space was $\Delta x = 2L$ (It could easily be millions of light years.) Now further suppose that there is a detector on a space craft just passing the earth at a speed v which also detects the two pulses. We assume that during the detections the velocity of the space craft is in the direction of the galaxy on the right and that the height above the earth of the space craft is negligible compared to L so that we can assume that the detector on the space craft was at essentially exactly the same point in space as the earth when the pulses were detected. Under those conditions we can use the Lorentz transformations to determine the space and time separations of the two pulse emissions from the galaxies as inferred from the data collected on the space craft. Referring to quantites determined from the space craft as primed we have

$$\Delta x' = G(2L)$$

and

$$\Delta t' = G(-(v/c^2)(2L)) = -2GLv/c^2$$

because in the unprimed frame on earth the time between the events was observed to be zero. The inferred spatial separation has changed due to the factor G. What is more remarkable is that the inferred time separation is not zero. From the data on the space craft one infers that the pulses were emitted at different times with the one on the right occurring first. That is not because the detector on the space craft was closer to the galaxy on the right. We have assumed that the earth and the space craft were coincident, that is in the same place, when the detections were made. So two events which appeared to be simultaneous in one frame (the earth's) do not appear to be simultaneous in another (the frame of the space craft containing a detector). But it gets weirder: Imagine another space craft going by the earth in the opposite direction and at the same speed but opposite velocity and also with a detector. It also detects the two pulses. One infers the same space separation from that data but yet another time separation:

$$\Delta t'' = +2GLv/c^2$$

that is the data taken on the pulse on the left occurred first. This actually plays havoc with Newton's concept of time and of common notions of the past and the present. If we think of an ancient 'present' when the earth observers inferred that the pulses were simultaneously emitted then one infers from the first space craft detector that the pulse from the right galaxy was emitted before that present , that is in its past, and that the pulse from the left was emitted after that present, that is in the future of that present. From the data on the second spacecraft one infers the opposite conclusion, that the pulse from the left occurred in the past and the one from the right in the future. That is, not only have the rates of the clocks changed. The very order of events and the definitions of past, present and future, depend on the reference frame from which the data is collected. (The effect in the example is not extremely small: with a typical space craft velocity of $v/c = 10^{-5}$ and a typical L of 10^7 light years the magnitude of $\Delta t'$ would be $2 \times 10^2 = 200$ years.) Fortunately these effects only make the order of events indeterminant in cases in which the two events in question are so far apart in space that there is not time in any frame for a light pulse to go from one location to the other before the later event takes place. Such events are called 'space-like separated'.

5.5 PROPER TIME

Now that we have established that clocks change their rates and can even give confusing signals about the order of events in different frames of reference, one may ask whether any stable sense at all can be made of time in this realm of frames of reference which are rapidly moving with respect to one another. It turns out that it is possible to do so to a limited extent as I discuss next.

Given the space and time separations $\Delta x, \Delta t$ between a pair of events in some reference frame, we can calculate the quantity Δs^2 defined as

$$\Delta s^2 = \Delta t^2 - \Delta x^2/c^2$$

where, for this discussion, you can think of Δx as the spatial separation of two events along a line between the locations of the two events. To get some feeling for the meaning of this, we note that the units of Δs^2 are time squared. Also if the two events are the arrival of a light pulse going from one spatial point to another then $\Delta x = c\Delta t$ (because the light speed is the same in any frame) so that $\Delta s^2 = 0$. What is remarkable, and not very obvious, is that if you insert the Lorentz transformation equations into this definition you find out that Δs^2 is the same in every frame moving at a constant velocity with respect to another. The algebra to prove this is a little more complicated than the level that I am trying to maintain here so I put it in an Appendix 5.3. Thus by taking the square root of Δs^2 we might hope to find a universal way of determining and thinking about the time between two events. That works if $\Delta x^2/c^2 < \Delta t^2$ because then Δs^2 is a positive number and we can take the square root and get a number which has a natural interpretation as the time between the events. In that case we say the events are 'time like separated'. There is time for a light pulse to get from one of the events to the other before the second one occurs. In that timelike case, we call $\sqrt{\Delta s^2}$ the 'proper time' between the events. (Actually it's $\pm\sqrt{\Delta s^2}$, but if you have information about which event came first in a frame in which they both occurred at the same spatial location you can choose the correct sign.) The proper time is then a satisfactory way of describing a universal measure of the time between events as long as they are time like separated. The example of the decaying muon passing Echo Lake and then arriving in Denver which was discussed earlier falls in this category of two time like separated events. A feature to note is that there was one frame (the frame attached to the downward moving muon) in which the two events occurred at the same place ($\Delta x = 0$). That feature is typical of time like separated events. Any moving entity (some collection of matter, such as a particle or a person) will experience a series of events which will occur at the spatial origin of the frame moving with itself and those events will be correspondingly timelike separated. The events will occur, one after another in a frame independent order at the spatial origin of the frame moving with the entity and the proper time of each event relative to the first can be used as a measure of the progression of the entity through its history. This line of events is called a 'world line' for the entity. I think you can see that it is intuitively all right to think of your own history that way. What is different about the relativistic picture is that measurement like events which record the events occurring along your world line in frames moving with respect to you will not be recorded as all occurring at the same spatial position and though the events will occur in the same temporal order in those other frames, the magnitudes of the times between them will be altered, as they were for the muon.

If $\Delta x^2/c^2 > \Delta t^2$, this entire way of thinking about the meaning of Δs^2 doesn't work because Δs^2 is negative. In that case we say that the events are 'space-like separated' and there is not time for light to travel from one event to the other before the second event takes place. Since Δs^2 is < 0, taking the square root gives an imaginary number which has no obvious physical interpretation. However we can take the square root of $-\Delta s^2$ which in the space like case is positive and we get two real numbers, one positive and one negative, corresponding to the fact that such events can be recorded in different temporal orders in different frames. The two supernovae flashes described in the last section were an example of space like separated events. We saw there that the temporal order in which the events occurred, as well as the amount of time between them, depended on the frame in which the events were recorded.

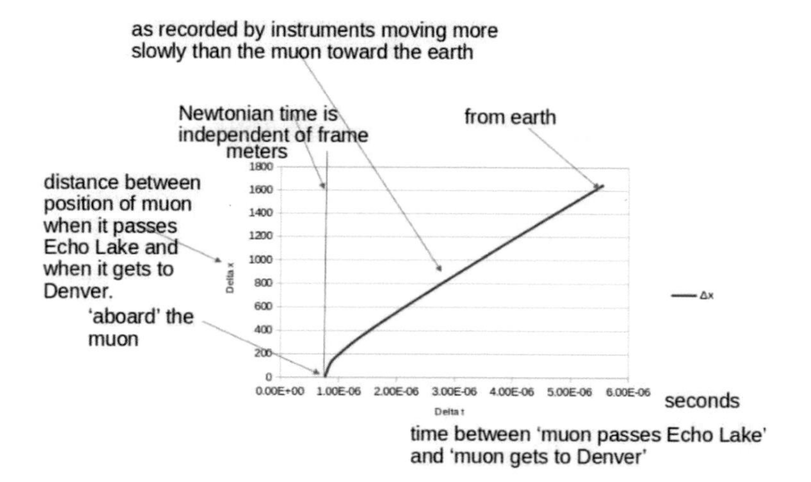

Figure 5.4 Space and time separations between the arrival of the muon at Echo Lake and its arrival at Denver from the viewpoint of various frames of reference. $\Delta x = 0$ corresponds to the frame of reference traveling with the muon and $\Delta x = 1642$ *meters* corresponds to a frame of reference fixed in the earth.

It is instructive the display the equations resulting from these four cases which relate $\Delta s = \pm\sqrt{\Delta t^2 - \Delta x^2/c^2}$ in the time like case and $\sqrt{-\Delta s^2} = \pm\sqrt{\Delta x^2/c^2 - \Delta t^2}$ in the space like case in graphs of Δx versus Δt as shown in Figures 5.3 and 5.4, bearing in mind that, for a given pair of events, Δs^2 is the same in all frames. In the first graph I show how the example of the muon appears in such a graph. The two events in question were the passage muon past Echo Lake and its arrival in Denver. The frame in which those two events occur at the same spatial point is one which is fastened to the muon. The curved line shows all the possible values of Δx and Δt which could be measured for the pair of events in different frames associated with different velocities at which the instrument recording the occurrence of the two events is moving with respect to the muon . Δt_0 is the time which passes in the frame of the muon

In the second graph I show how the events in the example of the two light flashes from supernovae in the last section appear on such a diagram. The two events are space like separated and it is easy to see from the graph that the time separation can be either positive or negative, consistent with the result which we got directly from the Lorentz transformation. Diagrams like this are called Minkowski diagrams, after the person who first suggested depicting events in space and time in this way.

Notice that if we are talking about events which correspond to the arrival of a pulse of light at two different places, then $\Delta s^2 = 0$ so the values of Δx and Δt are related by $\Delta t = \pm\Delta x/c$ which geometrically corresponds to straight lines in the Minkowski diagram. The pictures are all drawn for just one spatial dimension but we live in three, so we should really consider a four-dimensional space with three-spatial axes and one time axis. However that is hard for humans to visualize so people often

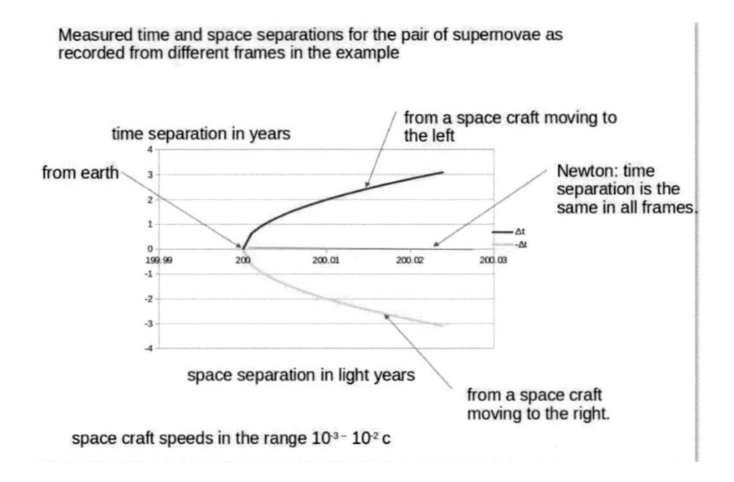

Figure 5.5 Time separations between the two supernovae as observed by the two space craft traveling at various speeds past the earth in opposite directions. The point on the Δx axis corresponding to $\Delta t = 0$ is the spatial separation as measured from earth in the example.

generalize to just two spatial dimensions and one time dimension so that they can draw pictures to visualize what's going on. Then the straight lines giving various ways of describing light propagation in different frames become 'cones' and a lot of popular literature refers to such 'light cones'. Notice that all possible time like separations occur on the side of the light line closest to the time axis and all possible space like separations occur on the side closest to the spatial axis. Very often these diagrams are rotated so that the time axis points up for the timelike case as well as for the spacelike one. In most daily life we deal with pairs of events which are close enough in space to one another so that light can easily get from one location to the other before the second event takes place, ie the events are time like separated and proper time gives a very satisfactory way to order the events and measure the time between them. The fact that the propagation of light between two points in space takes zero proper time is actually consistent with our intuitive impression that light propagates instantaneously, because our sensory equipment is not capable of resolving the very short times involved. When the Minkowski diagram is drawn on these everyday time and length scales, the light line is nearly vertical, leaving very little space for space like events and the corresponding anomalies. However there are plenty of examples in high energy physics experiments and in astronomy in which the time scales are shorter and the relativistic effects discussed here are dramatically manifest and observed in agreement with special relativity.

CHAPTER 6

Afterword

'If you knew Time as well as I do,' said the Hatter, 'you wouldn't talk about wasting IT. It's HIM.' 'I don't know what you mean,' said Alice. 'Of course you don't!' the Hatter said, tossing his head contemptuously. 'I dare say you never even spoke to Time!' Lewis Carroll, Alice's Adventures in Wonderland (1865)

6.1 INTRODUCTION

These questions were posed early in this narrative, partly following Augustine's famous essay, written almost 2000 years ago:

How is time measured and what does that tell us about its nature?

What is the nature of the past, present and future?

Why does time seem to have a direction and to be unstoppable?

Does time have a beginning?

Here I review some key points I have discussed with regard to each of these in the hope that it can help the reader develop some improved intuitions concerning the answers. It cannot be said in every case that definitive answers are available and there is a sense in which all of these questions are still open.

6.2 TIME MEASUREMENT

The review of clocks and time standards in the first chapter reveals that in practice apparently regular events observed in nature are used to set the time standard. The most important of those apparently regular events were, for many recent centuries, the movements of the planets and, in particular the time for the earth to complete a rotation about its axis. In the last 50 years, that standard has become secondary and the principal standards refer to regular motions which occur in atoms, notably the oscillations between states in the 133-Cs isotope of cesium. Secondary standards also include pendula in pendulum clocks and vibrations of atoms in crystals.

The first lesson from this history is that time is measured between any pair of observed events by some device which effectively determines the number of times (or fraction of a time) which one of these standards takes to complete its regular cycle. For example, when you look at your i-phone to determine the time, the device is using

DOI: 10.1201/9781003037125-6

the number of times a cesium atom has oscillated since last midnight to determine the time it reports to you. As I pointed out, the use of atoms in this way implicitly assumes several things. The oscillations in the cesium atoms are assumed to be the same in all the copies of atoms of that isotope of cesium and in all places and at all times. There is plenty of evidence that cesium atoms act identically during short periods. Slow drifts of their behavior over longer times would be much more difficult to discern and are usually assumed not to occur. There is also some evidence, from the consistency of evidence from laboratories around the world, that the location of the atoms does not affect their oscillatory behavior. However we saw that special relativity implies that, if the atoms are moving with high velocity with respect to one another, then the times between events will certainly be found to be different depending on which of the sets of atoms is used as a clock. I have not discussed the fact that Einstein's theory of general relativity predicts that large gravitational fields can also affect the readings from clocks in those fields. It turns out that both of those relativistic effects are signficant in the operation of the global positioning system (GPS) and the associated corrections are essential to its reliable operation.

The second lesson from the brief history of measurement is that most events which we observe in nature are not suitable for use as clocks. Most motions, even if they appear to be regularly oscillatory for brief periods, eventually slow down and stop. That is what happens to a pendulum, for example. The reason is fundamentally associated with the second law of thermodynamics: There is basically only one way to describe a perfectly regular motion at the atomic level. That implies that the entropy of a system moving with perfect regularity has zero entropy. But we observe that the entropy, evaluated at a coarse-grained scale, of nearly all systems containing large numbers of interacting atoms increases with time. For clock-like motions this manifests itself through interactions with the surroundings of the apparently regularly moving system, as in the friction of the bearings of a pendulum and its interaction with the air. Such couplings with the larger world are absolutely necessary to make a useful clock because one has to find a way to read it and that requires some interaction with a macroscopic phenomenon which humans can discern. The interactions have been minimized in the most useful clocks: The earth's rotation is slowing but it is rotating in the near vacuum of space and the rate of slowing is very small. By the design of atomic clocks the interactions with the larger world are minimized by observing the motions in vacuum and by use of very gentle readouts with microwaves and lasers. We have discussed how the effects were minimized in pendulum clocks by the use of escapements. The implication concerning the meaning of time is that the standard and thus the meaning of time is associated with idealized systems whose entropy remains zero as they oscillate. The idealization is a human choice, but it is very natural and can be seen historically to be associated with the utility of developing the ability to predict the changes of the seasons in order to grow crops and more recently to build systems, such a railroad networks, airline systems and the internet, which permit coordinated actions by large numbers of humans.

Comparing this story with Augustine's discussion of time measurement, one can see that he had some things right. (Yes this is a 'Whiggish' account of the history.) We measure time between events by starting to count the oscillations at the first event and stopping when the second event occurs. (That was discussed in some detail in the dock-canoe-water wave example in Chapter 5.) All the measurements are made in the present and all the records of the count are also made in the present. We do not go back in time to measure past events. We discuss and analyse the time that passed between the events using the present records which are in turn assumed not to have changed since the events took place. Thus for human utility there must also be a mechanism for stably storing the records. We discussed that in some detail with regard to the records of the collective human past in Chapter 2. The assumption of the stability of records remains a problematic one for human analysis of past events and that has come up several times in this narrative. On the other hand Augustine's ideas about techniques for measuring time seem quite naive to a modern reader and one can safely say that enormous progress has been made since Augustine's day with regard to the technological aspects of time measurement. Even there, one can see that Augustine was attempting to find a way to associate biological rhythms with time measurements and we have seen that recent progress shows that all organisms do indeed measure time using regular biochemical processes.

6.3 THE NATURE OF THE PAST, PRESENT AND FUTURE

In Chapter 2 the Newtonian, sometimes termed the Galilean perspective was described: Times can be described with numbers which can take any value in the real number system (numbers including any decimal fraction). One of those numbers can be used to describe the present and then all the events in the universe can be associated with a number by measurements of the time between those events and the present. All the events in the universe are supposed to experience the same time independent of whether oscillating systems used as clocks in them are moving with respect to one another or not. As discussed in Chapter 5, some of those assumptions are known not to be correct and the errors are especially large if the clock systems are moving with high speeds with respect to one another. However for most phenomena which humans experience from day to day this Newtonian way of thinking about time is adequate to quite a good quantitative precision. However the Newtonian perception of the present, which agrees with qualitative remarks by Augustine written centuries earlier, is quite strange and paradoxical to human intuition. Newton regarded the present as infinitesimally short. The difference is that Newton (and Leibniz at almost the same time) found a way to make sense and use of that notion in a way that permitted him to make precise predictions about the motions of material objects. That has been discussed in some detail in Chapter 2. We saw that the mathematics can be interpreted to mean that some useful properties of the present, namely the instantaneous velocities of the particles in it, are actually determined by the immediate past positions of those particles. Given those assumptions and techniques and a mathematical model of the forces between the particles, Newton made precise predictions of the motions of the planets around the sun which agreed spectacularly well with astronomical observations and convinced the scientific world of the validity of the Newtonian perspective. The mathematical and observational aspects of this perspective continued to be refined and to experience dramatic success for more than two centuries after Newton. Among other developments, electromagnetic forces, of which Newton knew nothing. were successfully incorporated into his framework. In recent years, the Newtonian perspective has been used to develop descriptions of biological systems at the atomic level and has been useful in understanding the dynamical properties of biomolecules.

However, regarding the past and the future, this Newtonian perspective, if applicable, as seemed increasingly possible, to all of nature, has some humanly disturbing consequences. Newtonian predictions are mathematically exact. If they are a full prediction it means that both the past and the future are fixed. The entire history and future of everything is following a fixed path through the space of the positions and velocities of the particles the universe contains. This raised, and, despite several new developments in physics, continues to raise questions about free will and causality. For example, we have cited the growing understanding from neurobiology of how organisms make choices. Most of the experiments have so far been on rodents but some results are available for humans as well and the choices are very simple ones. Nevertheless, it becomes quite easy to imagine that a detailed description, based essentially on Newtonian physics, of how more complex decisions are made will emerge. Since Newtonian physics is deterministic that might suggest that 'free will' is an illusion. Whether that is a correct perspective is currently debated. I suggested a perspective in Chapter 3 that straddles the issue but may be useful: As noted in Chapter 3, Newtonian trajectories of complex systems (including, certainly, biological organisms) are almost always 'chaotic' in the technical sense that tiny changes in the starting conditions can lead to enormous changes in the results at later times. If organisms are essentially Newtonian that would mean first that a choice, though predetermined by past conditions, might nevertheless have enormous consequences for the organism. So the Newtonian perspective does not suggest that choices don't matter. But were those choices predetermined? To get another view of that recall the notion of coarse graining and the related idea of 'emergence' discussed in Chapter 3. The notion of a human being is a 'coarse-grained' one: One does not describe a person by listing the positions of all the atoms in all the biomolecules in the person at each time. Even doctors don't do that. A much smaller set of variables is used (eg temperature, oxygen in blood, glucose in blood, heart rate, blood pressure) corresponding to a very coarse-grained description. If one confines ones attention to that coarse-grained scale, which is certainly unavoidable from day to day, then within that macroscopic description, when a person does one thing instead of another a choice, which could not be predetermined using those coarse-grained variables, has been made. We may say that 'free will' is an emergent property only apparent on a coarse-grained scale.

With regard to causality, the Newtonian perspective, if taken seriously, can also be disturbing: The path through the space of positions and velocities of all the particles in the Newtonian universe is fixed. If one knew the positions and velocities of the particles in the future one could mathematically follow that path back to the present. Thus, in a sense, future events can reasonably be regarded as causing the present ones, though the usual perspective is the opposite one, in which past events cause present ones. As mentioned several times earlier chapters, that fact crops up in many of the mathematical descriptions of Newtonian physics and the resulting 'acausal' results which would permit the determination of the present using data about the future are simply ignored. It appears that the reason that humans usually assume that the past, and not the future, determines the present, is that, through the available records, only data about the past is available and data about the future is not, so only the future can be predicted. However in a strictly Newtonian framework, present data could be used to postdict the past using Newtonian mathematics. Thus it appears that the direction in which 'causality' works. at least in a Newtonian context, is a convenient convention and not a law of nature.

When one takes into account the effects of quantum mechanics, which alter the Newtonian perspective on small scales, then views of the past present and future are not so definitively clear and depend to some degree on what interpretation of quantum phenomena you prefer. At the nonrelativistic level, there are several interpretations though which one you prefer has little or no effect on the successful use of quantum mechanics to predict and interpret the results of experiments. In the 'Bohmian' or 'pilot wave' perspective, one may regard the phenomena of the (nonrelativistic) universe as described by one of a set of trajectories through the space of positions of all the particles in it, in which each of the particles is moving at a velocity determined by the wave function. The set is densely packed and there is an infinite number of them, but the real system is moving along just one of them. Once it is known which of the trajectories is the one describing the universe, the future, as well as the past is determined, much as in Newtonian physics, though the trajectories are not the same as the Newtonian ones for the same starting conditions and are heavily influenced by the wave function. Unlike Newtonian physics, there is no way to determine the starting conditions precisely and they are selected at random by using a probabilitiy distribution determined by the wave function. The wave function, on the other hand is completely determined by the starting conditions and the dynamical equations of the theory. In this way, in the Bohmian interpretation, the past and future are fixed but cannot be predicted because of the random element in the selection of initial conditions. However in some experiments, if the wave function is known the trajectory can be postdicted after a measurement, thus determining the fixed past of the system. In the Bohmian or pilot wave interpretation, there are no special rules for interpreting measurement. It is assumed that measurements of positions can be made with arbitrary precision and, if such a measurement is made, one can learn which trajectory was randomly selected at the outset. Thus the history can be postdicted but not predicted. In another prominent interpretation, there is a somewhat similar description of the system. One considers a family of trajectories called 'histories' which describe possible pasts and futures though the mathematical description is quite different. The dramatic difference is that at least many, if not all adherents to this point of view want to regard all the possible trajectories as 'real' so that 'many worlds' or 'multiverses' exist. Since only the history which an observer is inhabiting can be observed, the supposed existence of the others makes no empirical difference and the two points of view are practically if not metaphysically, very similar. However the adherents to the latter view make use of 'Born rules' for describing measurements which are inherited from early formulations of quantum mechanics and are time irreversible, leading to a quantum mechanical 'arrow of time' which, unlike Newtonian or Bohmian mechanics makes it impossible to reconsruct the past from future data. Furthermore, in this latter school there is the view that positions of particles are only fixed at points in the history when the postulated measurements are made and that one cannot define, determine or measure the paths of the particles between measurements. So the nature of reality in that picture is a spotty one of definite positions of particles when measurements are made and only a wave function in between. The prominent role of 'measurements' in this latter set of ideas raises the issue of what, exactly, constitutes a measurement, because it cannot be assumed that what humans do in labs on earth determines the history of the universe. That issue is discussed in Chapter 4, with references to various points of view.

The takehome message from this discussion of the past present and future in a quantum mechanical world is that the issues are not settled but at least nonrelativistically a perspective consistent with the successful aspects of quantum theory is available which is time reversible, so that causality could run either way, as in Newtonian physics and in which the past present and future are fixed, not predictable but the past is, in principle, postdictable from present data. The alternative views are at greater variance with the Newtonian perspective. They are intrinsically time irreversible, postdiction is not possible and the reality of particle positions and velocities between measurements is regarded as a meaningless question. Strangely, these apparently quite diverse perspectives are consistent with what is currently known empirically from experiments on systems obeying quantum mechanics.

Relativistic physics, of which only special relativity was described in Chapter 5, describes reality in a way quite close to the Newtonian one in terms of the positions and velocities of point particles. Deferring comments on efforts to include relativistic insights into quantum mechanics, the positions and velocities are determined in a way quite close to the way it is done in Newtonian physics and the resulting trajectories are usually regarded as real. However, guided by the surprising observation that light propagates at the same speed in all reference frames, Einstein found that it was necessary that natural 'clocks' such as decaying nuclear particles and oscillating cesium atoms, run at different rates in different reference frames moving at high speeds with respect to one another. That appears to challenge the Newtonian idea of a universal time itself and it was shown in Chapter 5 that it even raises questions about whether a unique present, definable in all frames, exists. However it turns out, with some details described in Chapter 5, that as long as the pair of events in question occur closely enough in time so that a light pulse can propagate between them, then one can mathematically define a 'proper time' between the events which is the same in all frames of reference. In that case a 'present' exists associated with all the events occurring at the same proper time and the events of the past and future occur in the same order in all frames, though the length and (clock) time separations are different as recorded from different reference frames. When there is NOT time for a light pulse to propagate between two events then the events are said to be 'space like separated' and the conclusions are less comforting. Pairs of events that are space like separated can appear to be simultaneous in some frames and not in others and the order of events can appear to be different as recorded from different frames. In practical life, space like separated events are rarely observed because the very large value of the speed of light assures that light pulses can usually get from one event to the position of the other before it occurs. However in astronomy, space like separated events are quite commonly observed, and these effects must be taken into account. With the assumption that a later event can only be caused by an earlier one if there is time for a light pulse to get from the earlier to the later one, time-like separated events are often termed 'causally related' but this certainly does not mean that the earlier event necessary 'caused' the second one in the usual sense, but only that it could have caused it .

There are further changes in the perspective on time in general relativity which describes similar effects in the presence of large gravitational fields, near stars, for example. General relativity starts with a modification of the definition of proper time given in Chapter 5 and modifies it to take account of the effects of masses on the coefficients in that equation. It has had some remarkable successes in predicting, for example, the trajectories of light near massive objects and the signals obtained from black holes collapsing onto one another. But though it is the accepted theory of gravity, the number of observations which have successfully tested it is much smaller than it is for the other theoretical formulations involving time which are discussed here. For that and other reasons, including the fact that repeated efforts to fit general relativity into a quantum framework have failed, that there are some astronomical phenomena which may suggest a need for alterations to the theory and that the mathematical complexity of general relativity is daunting, I have chosen not to discuss it in much detail here.

Are the perspectives of quantum mechanics and special relativity reconcilable? One must say first, as noted earlier, that there is not presently a relativistic version of the Bohmian description of quantum mechanics so one can only deal with the majority perspective discussed above. In that context, the problem which arises with quantum mechanics as defined and discussed in Chapter 4 is that special relativity, when taking account of energies and momenta, reveals that particles can be naturally created and destroyed when their kinetic energies are comparable to their mass energies ($E = Mc^2$). That means that the number of particles is not fixed in time and a description with a fixed number of particles, used by both Newton and nonrelativistic quantum mechanics, will not work. Basically, the solution is to change the system description and describe the state of a system by an account of the number of particles together with their momenta (or their positions but not both because of the uncertainty principle) . Then 'events' are described as transitions between such states which may or may not change the number of particles. Thus nuclear decays such as those used in measuring deep time as described in Chapter 1 as well as nuclear burning in stars and reactions in high energy physics experiments can be described and predicted, given the right models for the physical interactions. What the right models are has been the major topic of high energy nuclear physics for the last 50 years and a lot of progress has been made. But though successful descriptions of many experiments as well as some successful predictions have been made, few new insights into the nature of time seem to have emerged. A clue may be lurking in one anomalous feature of those theories: They all predict that certain experimental quantities, such as the electron mass, are infinite. That would appear to be a show stopper but it turned out to be possible to use experimental values for those quantities and then use the resulting theory to make predictions of the results of experiments which have been confirmed, sometimes with spectacular accuracy. The infinities arise in the theories from its properties at very high energies which are related to very short times. For a while it was fashionable in the theoretical community to speculate that possibly that was because the theories assumed that time was continuous down to arbitrarily small scales and that perhaps it was actually discrete at some tiny scale. That is qualitatively related to speculations by Zeno made 2500 years ago. But there was no indication in the experiments of any such scale and attention has apparently shifted elsewhere in the high energy theory community, though some work in that direction continues [54]. (Actually one of the more attractive features of widely publicized string theories is that they are alleged not to have such infinities (but see [54]). However, despite a lot of hype, it appears that after more than 40 years of massive effort, string theory has been unsuccessful and cannot even describe known reality, much less make any successful predictions.)

Are the past, present and future real? As noted, supposing that some clear definition of 'real' in this context can be formulated, we can see that in view of the nonuniqueness of the definition of the present with regard to space-like separated events as implied by relativity, it is hard to make sense of an attribution of reality to the 'present' while contending that the 'past' and 'future' are not real. That is the main reason that most physicists prefer a 'block universe' perception of reality in which all past, future and present phenomena are equally real. It must be admitted that, for human experience, the 'local present' has a special status because we only usually collect sensory information about events in our brains and bodies during quite short periods of time which, though of finite duration, we regard as the 'present'. But that special (local) status does not seem to require a radical 'presentist' conclusion that the past and future are 'illusions'. Indeed, as we discussed, we store vast amounts of information about the past in our brains and collective records and continually make sometimes successful predictions about the future so it may be said that the past and future are very real in ordinary human perception, though the range of perception in both the past and future directions is limited, as it is not, at least in principle, in the physicists' 'block universe' picture. Think of riding a light rail train from the airport into a city you have not visited before. You sit at a window and new scenes appear and disappear in the window. Some you note as particularly interesting: A striking building for example. On your return to the airport by the same conveyance some days or weeks later, you sit by the window again. You pay little attention to most of the scenes passing by but suddenly you observe that the interesting building has been replaced by a pile of rubble (probably an instance of capitalistic 'creative destruction'). Though you were not thinking consciously about it you had retained an image of the building from the past in your brain and your brain had made a prediction that you would see it again. But if you did not go back to the airport in that way for a few years, you would have been less surprised and might not have even noticed what had happened. The human present in this example is contained (partly) in what you see in the window as the train moves along, but the past and future associated with those scenes are quite actively in your consciousness for some finite time. I should reiterate that both physicists (particularly cosmologists) and philosophers continue to vigorously argue in their journals about this old question of whether the past and future are real or not. (See for example [55] for an extensive, if quite biased, review without a lot of mathematics.)

Finally what about the human perspective on the past present and future? We have cited the growing knowledge of neurological processes which is starting to reveal how humans store information in their bodies and brains which permit them to recall events in the past and foretell events in the future. The descriptions are quite consistent with the early speculations by Augustine, though they remain very incomplete. We can anticipate that if scientific work on this topic continues at the present pace, then the improved understanding could lead to indications of how an individual could learn to control the capacity to recall and foretell events, thus becoming more effective in reaching goals and, perhaps, widening the personal present of the person. However we also noted that there are definite finite bounds to both memory and prediction and evidence indicates that many individuals do not even utilize the natural, untrained capacity that they are born with very effectively. The growing understanding of human mental processes could also be misused to manipulate and control people and that is a hazard which seems to merit watching. At the collective level the human past is recorded both intentionally in books, databases and cultures as well as in archaelogical records. I noted that the data is growing very rapidly in volume but shrinking almost as fast in longevity, so that we are accumulating much more information about an ever closer past while losing information about the deeper human past. We are in a sense storing more and more about less and less of the past. The volume of data threatens to overwhelm the ability of humans to make sense of it. Solutions using machine learning and 'AI' may yield practical results (some of them possibly pernicious) but little insight.

Regarding the collective human future we are currently facing the consequences of manifest human inability to think collectively and act coherently and effectively even when entirely predictable catastrophes lie in the future. As noted, that ability to plan seems incapable of effective collective planning into much more than a fifth of an average human lifetime as manifest in such phenomena as the discount rate, terms of mortgages, and the temporal length of political and educational programs. The fact that the main threats from climate change lay three or four decades in the future when they were first well understood in the 1970's seems to have paralyzed the human collectivity with regard to any effective action. It appears that this limited time scale may be inherited from evolution and very hard to overcome, but the survival of the species in the long run may depend on doing it. On the positive side, vastly increased data collection and storage capabilities provide, in principle, better tools for dealing which such global catastrophes, if the human will to use them effectively for long-term effect can be mustered.

6.4 WHY DOES TIME APPEAR TO HAVE A DIRECTION AND BE UNSTOPPABLE?

One of the ways in which it is alleged that time has a direction arises from the second law of thermodynamics, discussed in Chapter 3. Recall that it states that entropy always increases with increasing times. That appears to contradict what has been said about Newtonian and nonrelativistic quantum theories which are, at least in the Bohmian perspective on quantum mechanics, wholly time reversible. I claimed and tried to make clear in Chapter 3 that if one takes the coarse graining implied in the definition of entropy and the chaotic nature of Newtonian trajectories for large systems into account, then the time irreversibility of the second law and the time reversibility of Newtonian physics could be reconciled, provided that the universe started out in a low entropy state. That may help to resolve some theoretical puzzles, but when people ask the title question of this section, they are not necessarily or even usually worrying about the second law of thermodynamics or Newtonian physics. The perspective of the discussion of special relativity may be in that context more helpful. Note that in that discussion we focused on a description of reality through time as a series of events. The exact definition of 'events' was not specified, but in practice it was used to describe observations such as observations of the position of a wave. Attempting to think more generally we may think of each slice of Newtonian time in a Newtonian universe as a massive 'event' comprised of the appearance of all the matter and electromagnetic radiation in it. Some of the phenomena in that 'event' are traceable to similar phenomena in the previous slice but some are new, such as the appearance of a particle arising from a nuclear decay. If such a picture is accepted as a model for the real universe and the biosphere is included in it, then as discussed briefly earlier, a human being could be regarded as a set of events in the time slice. In the next time slice a similar set of events could be associated with the same human being but the event set will be somewhat different: a slightly different posture, some food digested, a different configuration of the heart and the ions carrying currents in the brain etc. In that perspective we are a sequence of events occurring throughout a life. All those events, in the 'block universe' perspective are real, but we perceive the present one. The perception itself is part of the present event. To put it another way, it is not so much that time is unstoppable as it is that the continual changes in ourselves are unstoppable (and largely dictated by our evolutionary heritage). It is not that we are either moving through time or watching it go by. We are IN time. That suggests regarding aging as a manifestation of the second law of thermodynamics. Regarding a human on a coarse-grained scale, the entropy starts out low and gets larger with time.

We discussed imaginary scenarios in which the universe starts to shrink, perhaps to eventually recur. Various cosmologists have concluded that even if the expansion stops and the galaxies start to come together again, that exact recurrence, such as that imagined by the ancient Stoics, is unlikely. Those speculative theories may not be correct because they all seem to make assumptions about the behavior of the universe which involve probabilistic selections of alternative phenomena at some point. If the trajectories were Newtonian or Bohmian then there would be no such probabilistic elements and the recurrence could be exact. However since no relativistic version of the Bohmian picture exists, we cannot be sure that there is any model leading to exact recurrence which is also consistent with known principles especially at the early and late stages of a recurrent universe. As noted earlier, there is no empirical astronomical evidence to support the hypothesis of a recurrence in the form of a 'Big Crunch' which might occur if the current expansion ended and length scales started to shrink. Nevertheless, some speculative theories posit that such shrinking will occur followed by a "Bounce" into a new period of expansion and even that such 'bouncing' has happened repeatedly in the past, with different physical laws in each recurrent universe. You may find such theoretical reasoning amusing to contemplate, but it falls into the category of ideas which, as explained in the Preface. are very weakly, at best, suggested by empirical evidence, do not much affect daily life and are not generally accepted by the scientific community. I have resolved not to discuss them further here and advise nonspecialist readers to regard them with appropriate caution.

6.5 DOES TIME HAVE A BEGINNING?

As noted in Chapter 1, the presently accepted cosmological model, massively supported by observational data at least in its later stages a few hundred thousand years and more after the initial 'Bang', can be dated to have occurred about 14 billion years before the present. If time began in some sense at the moment when the 'Bang' began then one might'say that time had a beginning. That is often stated in popular accounts and may be correct. However when we look back at what we mean by time it seems clear that something called time, much like what we consider time to be, will always be definable in a universe in which regular events occur. Did regular events occur before the Big Bang? Indeed, did any events at all occur before the Big Bang? Most of the models have nothing to say about that. The recurrent 'big crunch' models mentioned in the last subsection are an exception. and in recurrent models, one could certainly define time before the Big Bang.

That raises a related question which has unfortunately not been discussed at all here, namely does time have an end? Augustine does not discuss it, but in fact it was a part of the theology of his day that it does, as discussed in various ways in the Christian bible. I have briefly mentioned cosmological models, not favored by current evidence, in which the expansion of the universe is ultimately reversed resulting in some kind of 'big crunch' which might, or might not, be followed by a reexpansion (a kind of 'bounce'). More likely, on the basis of current evidence, is a scenario in which the expansion continues indefinitely, resulting finally in a very dead universe in which the stars are extinguished and the average matter density is very low. From the human perspective, of course, time ends for individuals in very short times. Whether it will end for collective human time as the species becomes extinct is of course unknown but likely given the observed history of species extinction which has already taken place and the many speculations concerning ways in which it could happen for our own species [6].

I am reminded of a story, possibly apocryphal, about former Vice President Al Gore. Unlike most presidents and vice-presidents, he seemed to have had a genuine interest in science, independent of its economic and political implications. The story is that he was meeting, on one occasion, with some eminent scientists to discuss such implications of recent scientific developments. As the meeting ended and the participants were leaving, Gore suddenly said "Wait, wait, what happened before the Big Bang?"

Appendix 1.1 Some Atomic Physics of the Cesium Clock

A cartoon sketch indicating what happens to the cesium atoms as they ascend and then descend in response to the downward pull of the earth after they are tossed up by the lasers is indicated in Figure 1.1A:

Assuming for the moment that they can be in two different 'levels' or I will say 'states' as implied by the international definition of the second quoted in Chapter 1, the atoms are all placed in the state which is lower in energy as they pass through the lower cavity on the way up and are started oscillating between the two states by microwaves from the second cavity as they pass, the first time, through the upper microwave cavity on the way up. The objective of the device is to tune the frequency of the microwaves emitted by the second cavity so that It is an exact integer fraction of the rate at which the cesium atoms oscillate between the two states. Once that is done, the microwaves can be (and are) used to drive a clock of a more conventional sort. If that tuning has been achieved successfully then when the cesium pass again through the upper cavity on the way down, they are all tossed into the upper state, they stop oscillating and the light from the probe laser (shown in orange in Figure 1.1) can be scattered from them. The scattered light hits the detector in that figure, indicating successful tuning has been achieved.

To make better sense of this explanation, it may be helpful to briefly review the meaning of some of the terms: Both light and microwaves, which are involved, are electromagnetic waves, consisting of electric and magnetic fields oscillating in space and time as schematically indicated in the next figure and in the animation shown in the accompanying web link. They propagate with the same speed in vacuum ($c = 3 \times 10^{10} meters/second$) but differ dramatically in wavelength (the distance between the peaks, traditionally called λ) . The light in the lasers which suspend the atoms in the cesium clock has a wavelength of approximately 8.52×10^{-5} cm whereas the microwaves have a wavelength of about 3.24 cm. That means that the peaks in the light waves are passing by a stationary observer more than 100,000 times faster than those of the microwaves. The microwaves are correspondingly easier to use in a electrical circuits to drive a clock and that is one of the reasons for translating the cesium frequency into the frequency of microwaves. (Microwaves have many other technological applications, including the transmission of signals to and from your i-phone.)

The two 'hyperfine levels' in the definition of the second refer to arrangements of the way the nucleus and the most weakly bound electron of the cesium atoms are

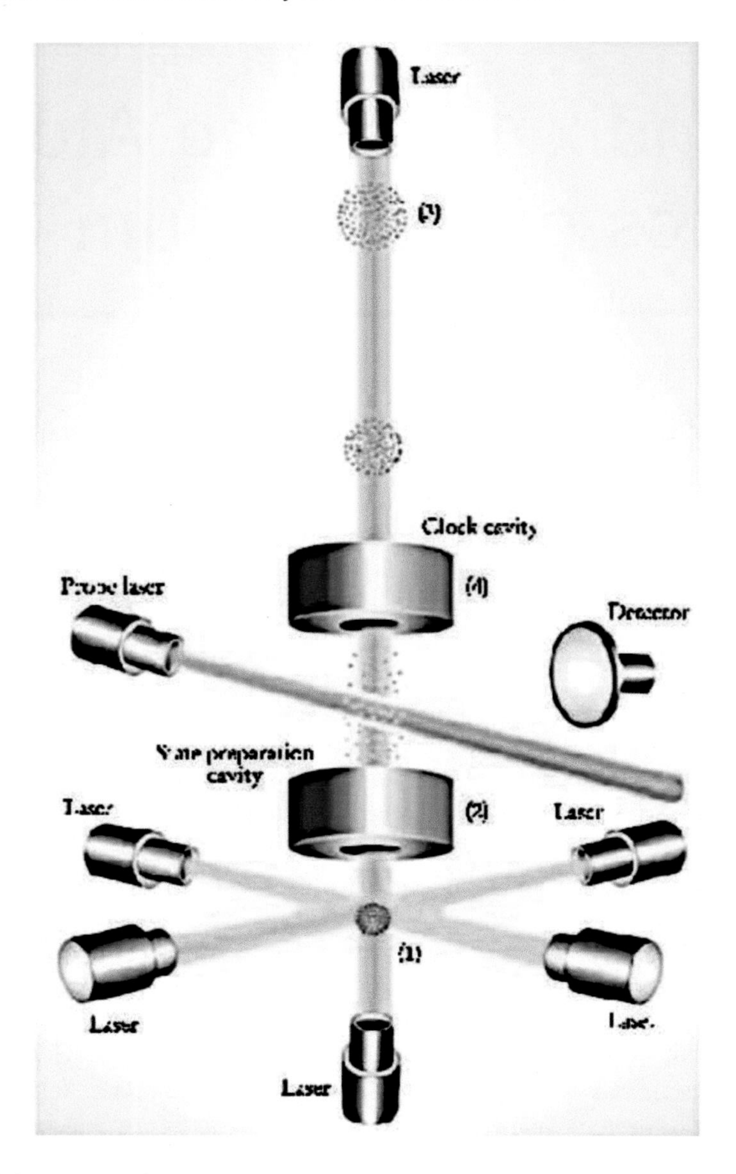

Figure 1.1A Schematic of the operation of the cesium clock. Reprinted courtesy of NIST, all rights reserved.

related to one another. To be clear about that, we need to be aware that the '133' in the phrase Cs-133 is the number of protons plus the number of neutrons in the Cs atom. The number of protons in the nuclei of all Cs atoms is 55 so there are 78 neutrons in the nuclei of the Cs atoms used in the clock. Since cesium atoms exist with other numbers of neutrons in their nuclei (but not different numbers of protons) we call Cs-133 an isotope of cesium. This cesium nucleus is distinguished from many of the others known by the fact that it is STABLE, meaning that it does not spontaneously turn into other nuclei and particles as time passes. (Unstable nuclei which do 'decay' in that way are used for deep time measurements as discussed in

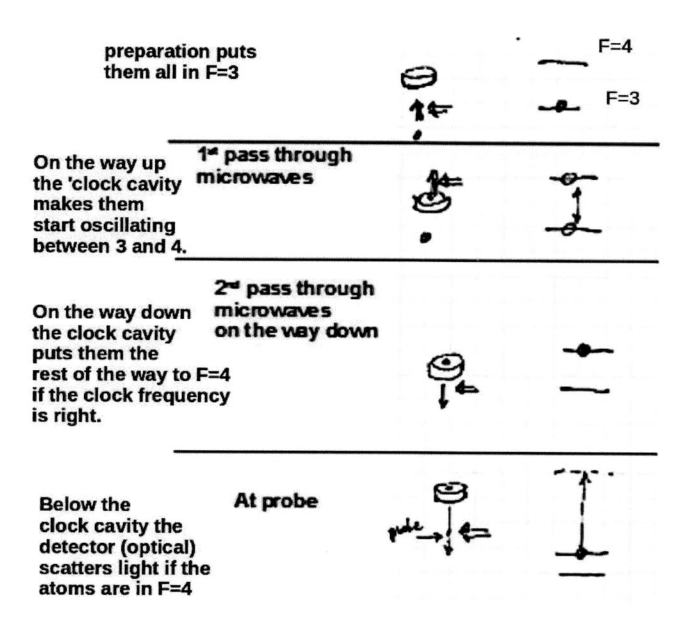

Figure 1.2A Another picture showing what happens to the states of the cesium atoms during the operation of the clock.

Section 1.6 of Chapter 1.) Another feature of the isotope Cs-133 is that the nucleus is spinning. It is a feature of nuclei, not shared by the earth and toy tops, that the spin is 'quantized' which means that the rate at which it is spinning can only have certain discrete values. Cs-133's spin, in units which will not be discussed here, is 7/2. (If you really wanted to know, the spin values are in units proportional to Planck's constant, which in turn is a constant of nature characterizing quantum phenomena.) The electrons in a cloud around the cesium nucleus are all spinning too, and the spin of each of them in the same units is 1/2. Finally you need to be aware that, in the cloud of 55 electrons surrounding the Cs nucleus, one is not very tightly bound to the nucleus and is easily caused to become detached or to move more rapidly in the cloud by tickling it with electromagnetic waves . Now we can qualitatively explain what is meant by the hyperfine levels in the definition of the second: It is another feature of quantum physics that, if they are interacting, two spinning particles can only be in stable (dynamically unchanging) states if the directions of the spins differ by certain discrete angles. This rule applies to the spin of the Cs nucleus and its most weakly bound electron, which must be either exactly parallel or exactly opposite in direction in order to form a dynamically stable state of the atom. Those two states (parallel and antiparallel spins) are the 'hyperfine levels' to which the definition of the second refers. They differ slightly in energy as indicated the rightmost column in Figure 1.2A. What happens in the clock is that, as a cesium atom passes the first time (going up) through the bottom cavity the microwaves force the electron spin to be antiparallel to the nuclear spin. The resulting total spin is $7/2$-$1/2 = 3$, labeled as $F = 3$ in the diagram. When passing the first time (going up) through the upper

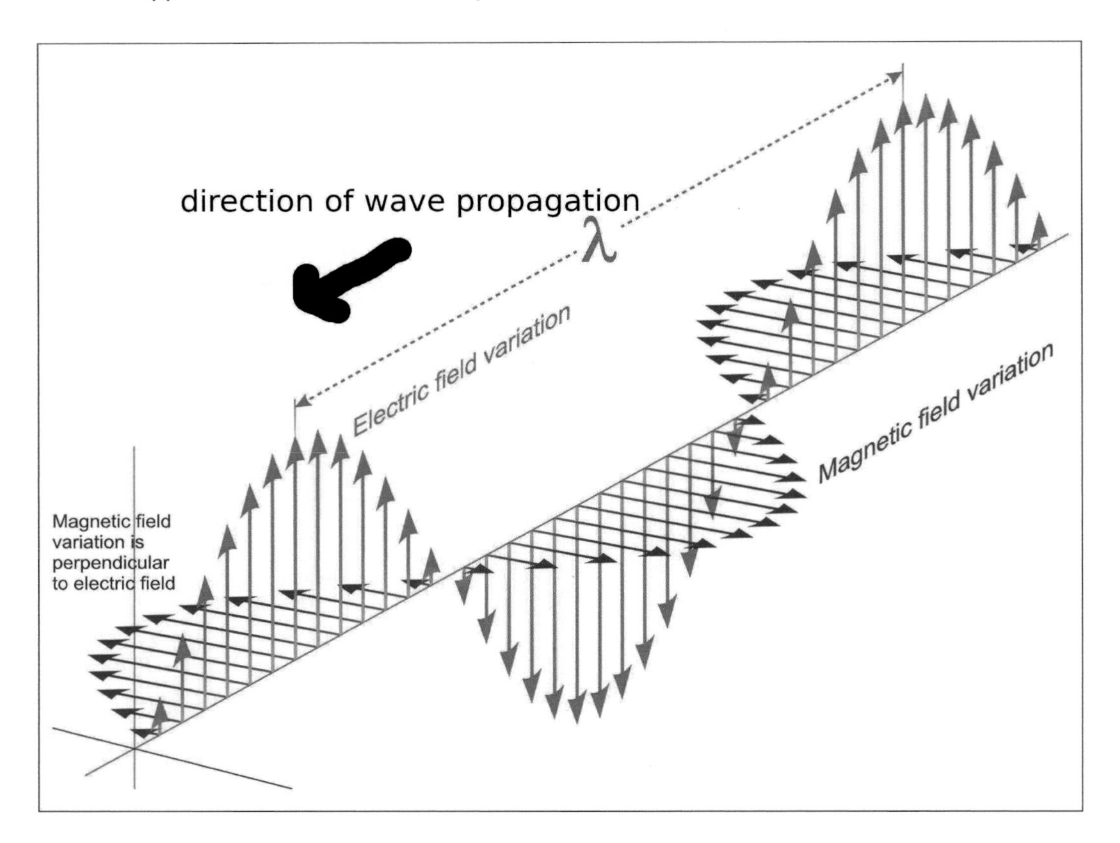

Figure 1.3A Schematic illustration of the electric and magnetic fields in an electromagnetic wave. For an animation see https://www.youtube.com/watch?v=GxJWrm4VNCA.

cavity, the microwaves interact with the electron again and start to turn it up into the parallel configuration with the nuclear spin, but they are only acting long enough to take the electron half way to the parallel configuration and, as the cesium atoms leave the upper cavity on the way up, they are in a dynamically unstable state in which the electrons spin is flipping back and forth between the parallel and the antiparallel configuration AT THE CLOCK FREQUENCY as defined in the international agreement. Now the atoms continue up the tube shown in Figure 1.1 and then back down again with each electron spin flipping up and down with respect to the nuclear spin the whole time (millions of flips). Meanwhile in the upper microwave cavity, the microwaves have continued to oscillate with the same frequency. When the atoms get back down to that cavity the second time they might be momentarily in the antiparallel configuration at a moment when the microwaves were at a stage at which they were pushing the atoms toward the antiparallel direction. If so, then the second exposure to the upper cavity does not push the electron spins all the way up into the stable parallel configuration and the detector shows nothing because the orange light of the probe laser cannot detect atoms with spins in the antiparallel direction. On the other hand, IF THE CLOCK FREQUENCY IS AN EXACT INTEGER

MULTIPLE OF THE FREQUENCY OF THE MICROWAVES IN THE UPPER MICROWAVE CAVITY then on the second pass of the cesium cloud on the way down through the upper cavity, the microwaves 'finish the job' of pushing the electron spins into the stable parallel configuration with the nuclear spin and the orange light of the probe laser is scattered off into the detector. In operating the clock the frequency of the microwaves is adjusted until the probe laser indicates that the frequency matching has occurred and then the microwaves of the upper cavity can be used to operate clocks of more conventional sorts whose time keeping is precisely synchronous with the international standard.

Appendix 1.2 A Few Facts about Molecular Biology

Here we review some basic facts needed to comprehend what happens in the biological circadian (daily) clock of the fruit fly as briefly described in Chapter 1 as well as the later description of what is known about memory formation and decision making in the discussion of the human past in Chapter 2. As noted there, the human circadian clock is believed to be similar to the one found in fruit flies, but it is understood in less detail.

A first, and remarkable, fact is that the biochemistry of all life in the terrestrial biosphere is extremely similar, strongly suggesting that all life arose from a common ancient ancestral system, whose detailed nature is unknown. As it works in the present era, the dominant molecular elements involved in living processes are nucleic acids and proteins together with some structural components including the lipids forming cell walls and carbohydrates used for some forms of energy storage which I will not discuss. It is well to bear in mind the length scales associated with the relevant molecules. The nucleic acids and proteins are long, linear molecules, constructed like chemical beads on a string, involving 10's up to many millions of individual atoms. As a result the characteristic sizes of these molecules (which are almost always folded up somewhat like balls of string and not stretched out like spaghetti) are of the order of a few millionths of a meter in size, whereas the individual atoms are characteristically less that a billionth of a meter in size.

I start with a description of the nucleic acids, popularly referenced by their acronyms DNA and RNA which stand for Deoxyribonucleic acid and Ribonucleic acid respectively. Since the 1950's DNA has been understood to contain a code for sequences of the biochemical entities in proteins and to be copied into new organisms during reproduction. For this reason it is often called an information carrying molecule and in popular imagination as well as some scientific practice to be assumed to control the nature, behavior and health of the organism bearing a particular version of the code in its DNA. I will argue that that interpretation is somewhat misleading and that it is closer to the mark to describe DNA as the library of the cell that contains it. A library is useful to users but it does not fully determine the behavior and character of the individuals that use it though it may limit them if the information in it is inadequate and harm them if the information is incorrect.

The 'beads' in the chemical strings which comprise DNA are called nucleotides, which are not to be confused with the nuclei of the atoms in them. The nuclei of the atoms in the nucleotides are more than 100,000 times smaller than the nucleotides.

Because of this difference in length scales, biochemists usually ignore the internal structure of the constituent atoms in the nucleotides and represent the atoms by colored spheres, tied together by sticks which represent clouds of electrons shared between adjacent atoms and tying them firmly together in what are called covalent bonds. The atoms, whose outer structure is a cloud of electrons which is usually not actually exactly spherical in nature, are colored in such a chemist's representation as given in Figure 1.4A, to distinguish the different atoms which each of the spheres represents. The colors have nothing to do with the actual colors of materials containing those elements so the colors should be regarded as just a labeling device. There are four different kinds of nucleotides used in DNA to form its codes. Their atomic structure is illustrated using the conventions just described in Figure 1.4A.

These elements are tied together in long strings to form DNA by phosphate groups consisting of a phosphorous atom and four oxygen atoms as shown in next Figure 1.5A. In this figure, the famous 'double helix' geometry of the molecule as it exists in living systems has been conceptually 'unwound' so that you can see how it is chemically stitched together. The double helix structure in which two DNA molecules are wound together as they often are is shown in Figure 1.6A

Part of the code in the DNA (and essentially all of it in organisms without (biological) nuclei at the center of their cells (these include bacteria and another class called archaea), is used, as stated above, to code for sequences of chemical elements in the molecular strings called proteins. (In general, a large molecule put together in a string like manner is called a polymer. Most plastics consist of nonbiological polymers and proteins, RNA and DNA are called biopolymers.) The 'beads' in the protein polymers consist of so called amino acids. The basic unit in the beads is simpler than the one in the nucleic acids and its structure is shown in Figure 1.7A. 'amino' refers to the presence of nitrogen in the structure. The 'R' in the figure refers to one of hundreds of chemical groups which chemists have discovered among the amino acids. The way that the amino acids bind together is that, when bonding, a water molecule is formed by taking an H off the N end of the first amino acid and an OH off the C on the other end of the second amino acid and allowing the C to bond to the N in a so called peptide bond. (Short protein strings are sometimes called polypeptides.) In the biosphere, 20 different R groups are used in the amino acids forming the proteins. The large number of R groups allows for an enormous variety in the kinds of possible proteins. The sequence of amino acid types in a given protein determines its function in living cells and those functions are correspondingly diverse. Proteins form the skeletal structures inside cells, act as valves in the cell walls, and speed up all kinds of essential reactions in the cell, both between other proteins and between other chemical constituents of the cell. (The latter function is called catalysis and the class of proteins that perform it are called enzymes.) That is only a partial list. It is fair to say that the proteins do most of the work of the cell. The atomic structures of R groups in the 20 amino acids used in our biosphere are shown in the Figure 1.8A. A table indicating how each amino acid is coded by 3 nucleotides in the DNA double helix appears in Figure 1.8A.

One more essential fact is required to understand the basic mechanism of the circadian clock in fruit flies: I mentioned that, in bacteria, the function of the DNA

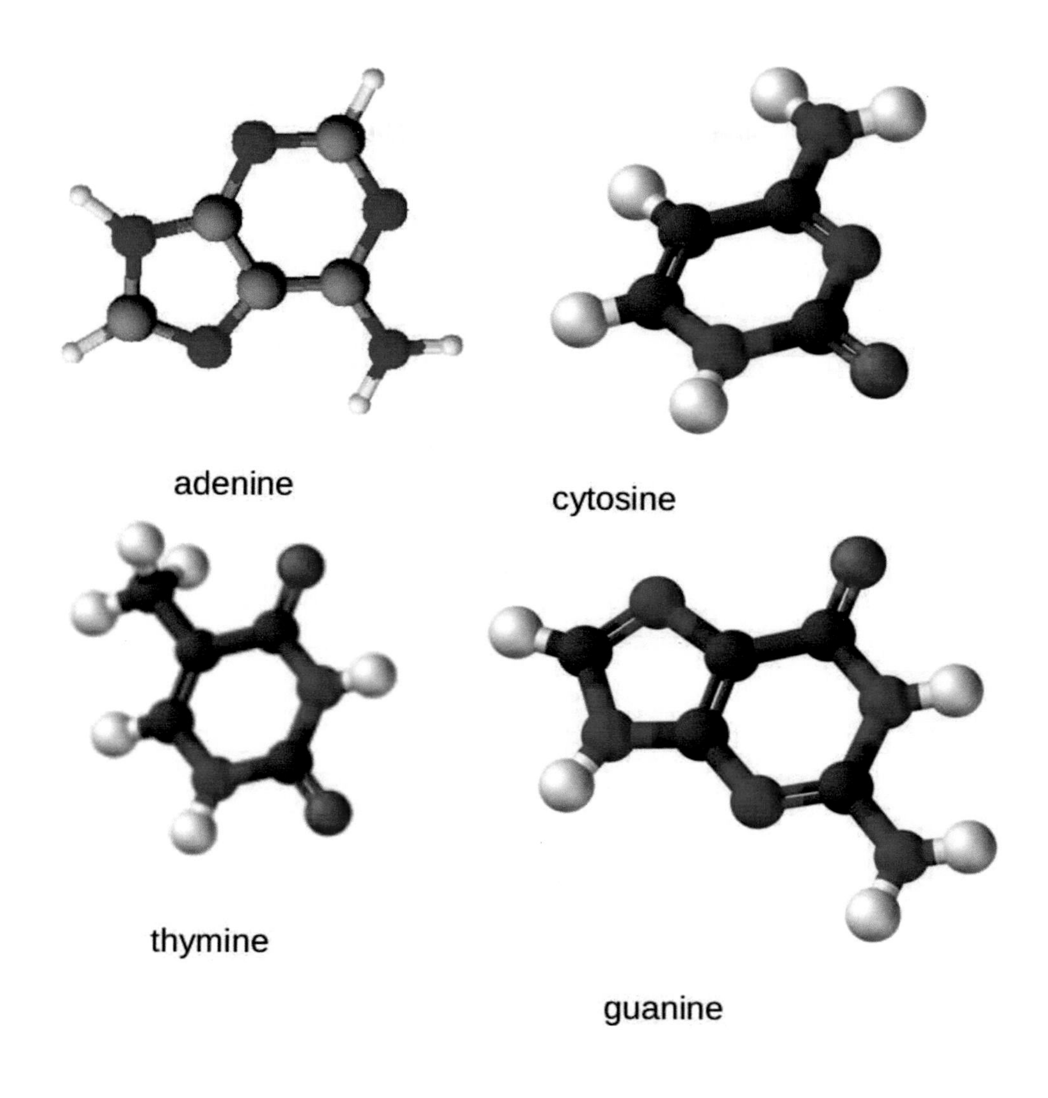

Figure 1.4A The atomic structure of the four nucleotides used by DNA as an 'alphabet' in encoding the sequences of amino acids which form proteins. The gray spheres indicate the location of carbon atoms, the blue spheres show the positions of nitrogen atoms, red spheres represent oxygen and white spheres represent hydrogen. The nucleotides themselves are called adenine (A), thymine (T), guanine (G) ahd cytosine (C). The 'sticks' in the figures represent strong covalent bonds which are usually stable at ordinary biological conditions. Not shown (see next figure) are weaker 'hydrogen bonds' which can bind A to T and G to C but not A to C or G or G to A or T. The preferential hydrogen bonding turns out to be important in the operation and translation of the code. The 'alphabet' of RNA is almost the same, but the thymine nucleotide is replace by uracil, denoted U, which differs from thymine only in the replacement of the single hydrogen sticking out to the right in the figure with a methyl group consisting of a carbon with three hydrogen atoms attached.

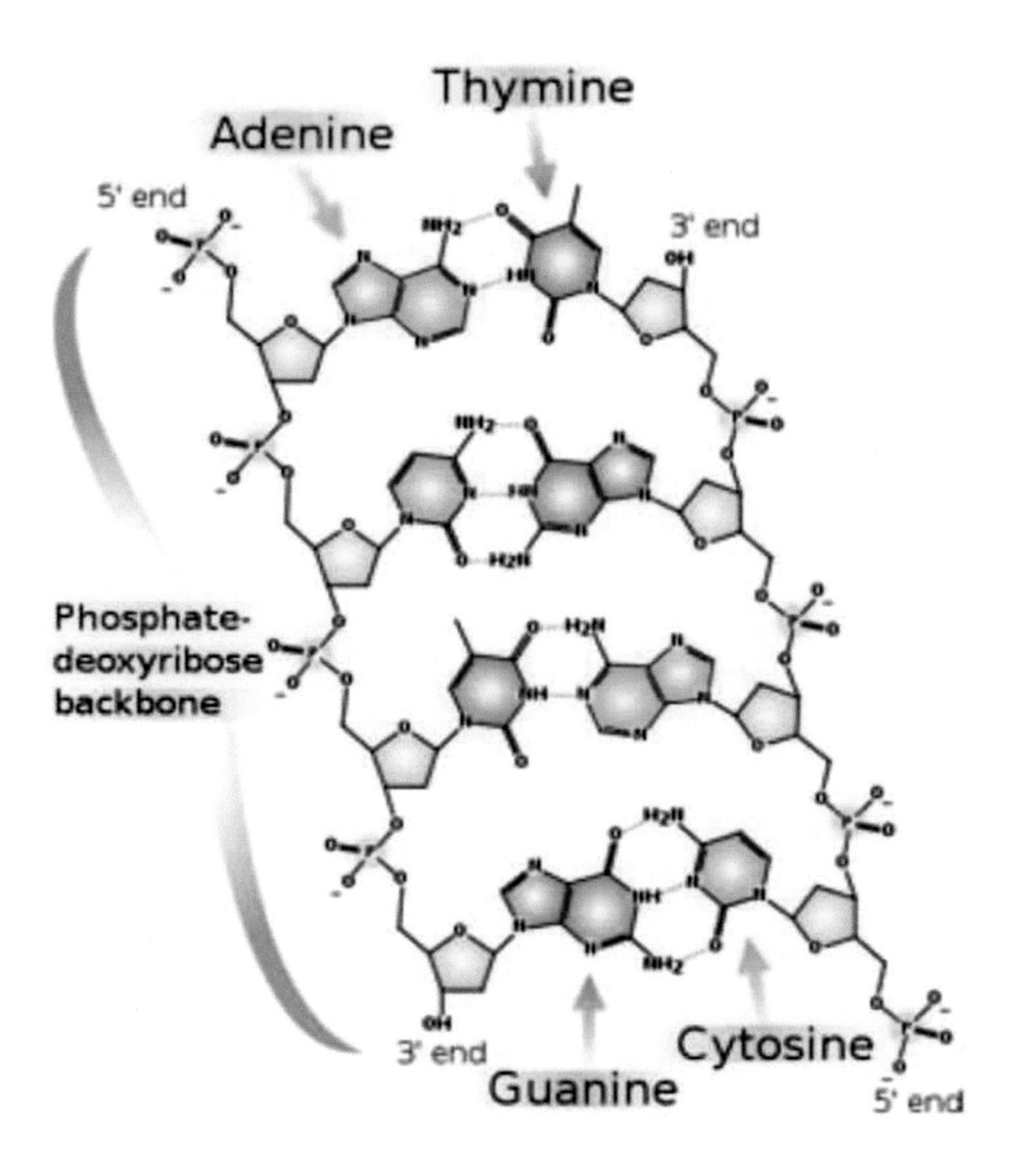

Figure 1.5A The atomic structure of a small piece of DNA, 'unwound' to illustrate the way the strings of nucleotides are stitched together with phosphate groups and the hydrogen bonding attaching one 'strand' to the other. When wound into a helix as shown in the next figure, it is called a 'double helix' because these two strands are wound around one another. The orange, five-ided structures containing oxygen and carbon atoms are called ribose groups. Ribose is a kind of sugar. The ribose accounts for the reference to 'ribo' in the full name for DNA. Figure by Madeleine Price Ball, in the public domain.

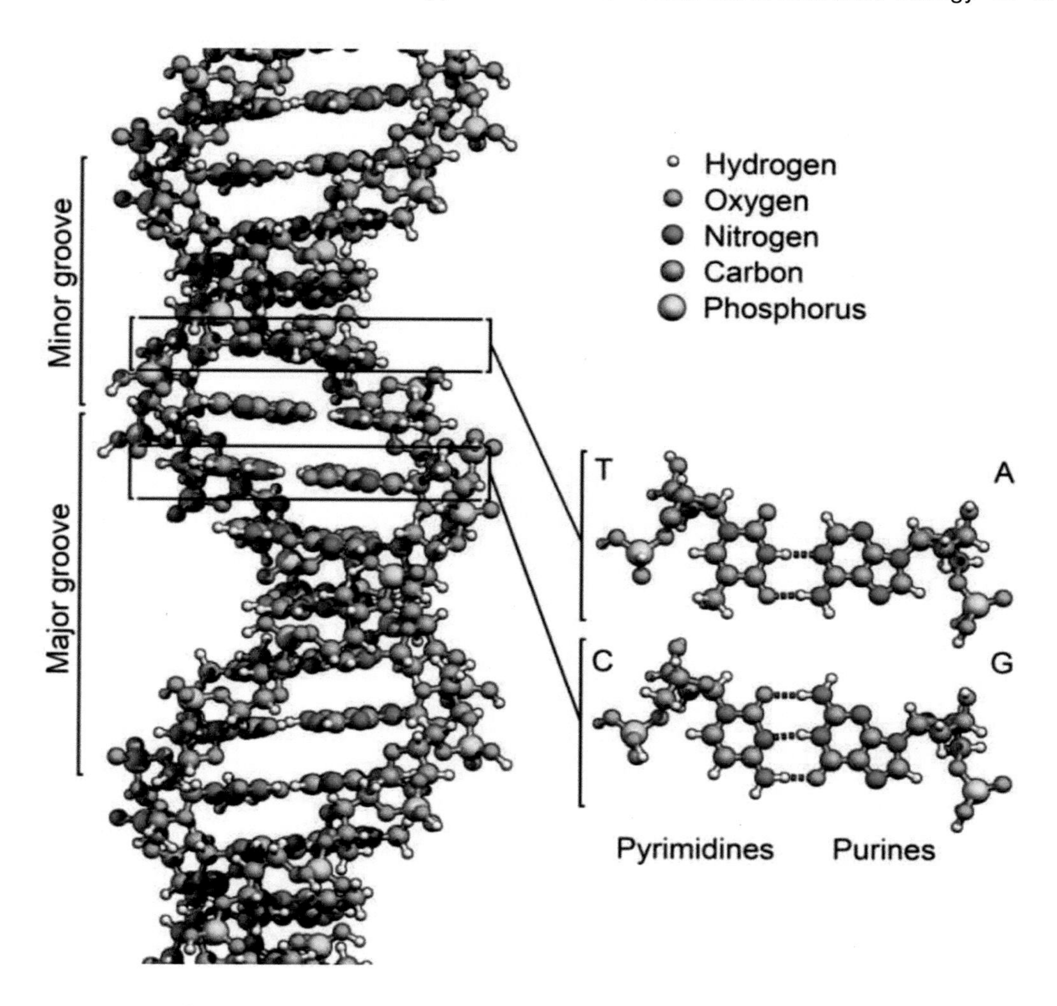

Figure 1.6A The atomic structure of DNA, when wound, to illustrate the way the strings of nucleotides are wrapped around each other and motivating the 'double helix' name. Figure by Richard Wheeler, used with permission.

code is almost entirely to code for proteins. Bacteria are unicellular organisms so the same set of proteins are always needed in their (single) cells. However after nearly 2 billion years of evolution, another form of life, termed eukaryotic, emerged on earth first in unicellular form, in which the DNA was bundled into a nucleus in the center of the cell. (Again we caution that the nucleus in question here is enormously larger than the nucleus of an atom and has an entirely different composition and function.) After a relatively short evolutionary time (around 10^7 to 10^8 years) multicellular eukaryotic organisms appeared. A fundamental difference with previous forms of life was that, in multicellular organisms, different cells performed different functions, EVEN THOUGH THE DNA WAS IDENTICALLY THE SAME IN ALL OF THE CELLS of the organism. How, for example, does a cell in one's nose perform a nose like function, while a cell on your toe or in your heart performs very different ones, when all those cells carry the same DNA? Understanding of the answer has come

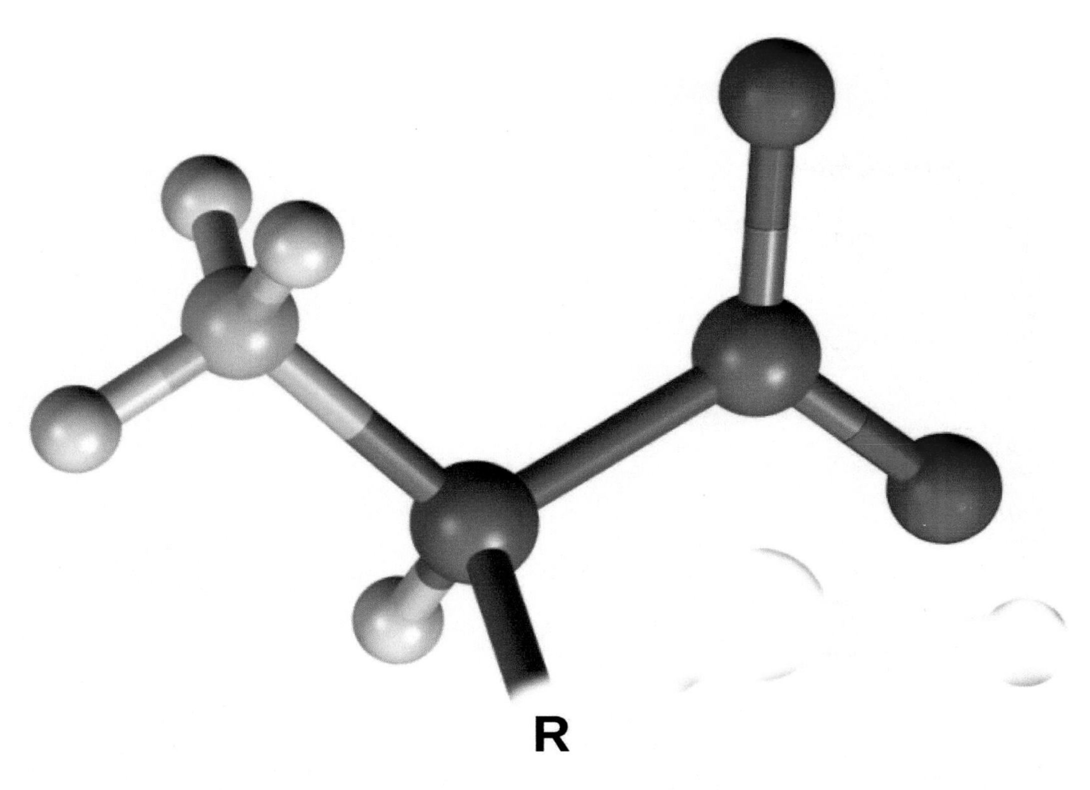

Figure 1.7A The structure of an amino acid, the basic 'building block' of proteins. The various possible 'R' groups used in the proteins of the biosphere are shown in Figure 1.8A. Adapted from image by Benjah-bmm27, Wikimedia Commons, public domain (https://commons.wikimedia.org/wiki/File:Glycine-zwitterion-from-xtal-3D-bs-17.png).

rather recently, When the codes in the DNA of multicellular organsims began to be determined around 2000, scientists found to their initial surprise that not all of the DNA was coding for proteins (as it had been found to do in bacteria and other prokaryotes.) For some time the parts of the DNA not coding for protein were termed 'junk'. In fact it turned out that the noncoding regions of the multicellular DNA were landing (or more traditionally binding) sites at which a protein could attach and thereby initiate the production of another kind of protein from a coding region nearby on the string. It is as if the landing of the protein in the noncoding region acts like a switch which 'turns on' production of a protein for which the code is nearby along the DNA. Thus a protein could land on one part of the DNA and initiate production of a protein needed for the operation of a cell in your nose, or, by activating a different part of the DNA could produce proteins needed by cells in your heart. It is as if the protein system in the cell determines a need and then sends a protein to the DNA to turn on production of what is needed. That kind of switching plays a role in the operation of the circadian clock in the fruit fly as described in Chapter 1. (It is also possible for such a switch to operate so that attachment of a

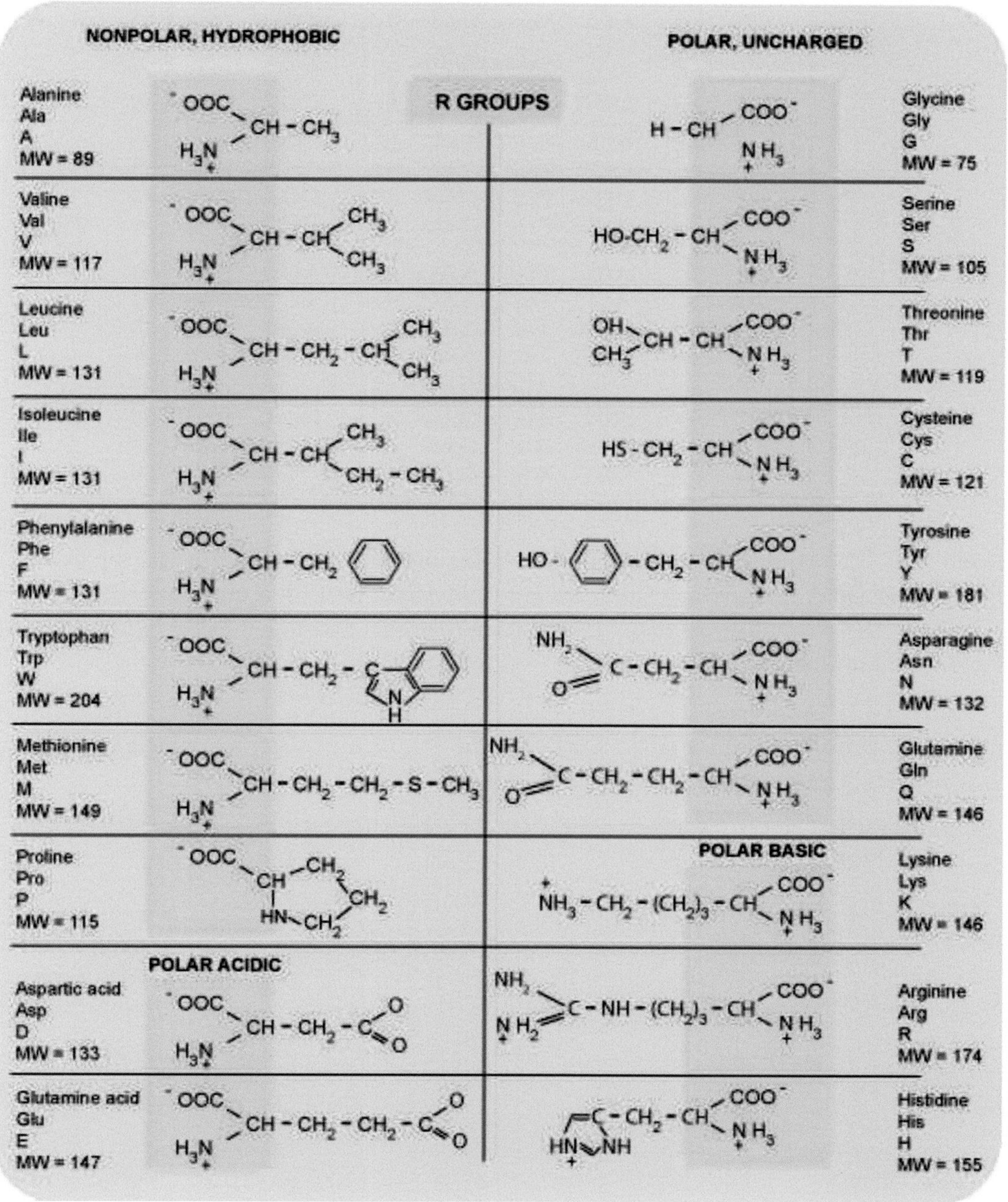

NONPOLAR, HYDROPHOBIC

POLAR, UNCHARGED

R GROUPS

Amino acid	Structure	Structure	Amino acid
Alanine Ala A MW = 89	^-OOC, H_3N^+ >CH - CH_3	H - CH< COO^-, NH_3^+	Glycine Gly G MW = 75
Valine Val V MW = 117	^-OOC, H_3N^+ >CH - CH< CH_3, CH_3	HO-CH_2 - CH< COO^-, NH_3^+	Serine Ser S MW = 105
Leucine Leu L MW = 131	^-OOC, H_3N^+ >CH - CH_2 - CH< CH_3, CH_3	OH, CH_3 >CH - CH< COO^-, NH_3^+	Threonine Thr T MW = 119
Isoleucine Ile I MW = 131	^-OOC, H_3N^+ >CH - CH< CH_3, CH_2 - CH_3	HS - CH_2 - CH< COO^-, NH_3^+	Cysteine Cys C MW = 121
Phenylalanine Phe F MW = 131	^-OOC, H_3N^+ >CH - CH_2 - (benzene ring)	HO - (benzene ring) - CH_2 - CH< COO^-, NH_3^+	Tyrosine Tyr Y MW = 181
Tryptophan Trp W MW = 204	^-OOC, H_3N^+ >CH - CH_2 - C (indole ring, N H)	NH_2, O= >C - CH_2 - CH< COO^-, NH_3^+	Asparagine Asn N MW = 132
Methionine Met M MW = 149	^-OOC, H_3N^+ >CH - CH_2 - CH_2 - S - CH_3	NH_2, O= >C - CH_2 - CH_2 - CH< COO^-, NH_3^+	Glutamine Gln Q MW = 146
Proline Pro P MW = 115	^-OOC - CH (ring: CH_2, CH_2, CH_2, HN)	POLAR BASIC $^+NH_3$ - CH_2 - $(CH_2)_3$ - CH< COO^-, NH_3^+	Lysine Lys K MW = 146
POLAR ACIDIC Aspartic acid Asp D MW = 133	^-OOC, H_3N^+ >CH - CH_2 - C< O, O	NH_2, NH_2^+ >C - NH - $(CH_2)_3$ - CH< COO^-, NH_3^+	Arginine Arg R MW = 174
Glutamine acid Glu E MW = 147	^-OOC, H_3N^+ >CH - CH_2 - CH_2 - C< O, O	(imidazole ring HN, NH$^+$) C - CH_2 - CH< COO^-, NH_3^+	Histidine His H MW = 155

Figure 1.8A The 20 ‘R’ groups used in the proteins of the biosphere. The pink part of each figure is another representation of the structure shown in the previous figure. Reprinted from www.neb.com (2021) with permission from New England Biolabs Inc.

First Letter	Second letter: U		C		A		G		Third Letter
U	UUU	Phe	UCU	Ser	UAU	Tyr	UGU	Cys	U
	UUC	Phe	UCC	Ser	UAC	Tyr	UGC	Cys	C
	UUA	Leu	UCA	Ser	UAA	Stop	UGA	Stop	G
	UUG	Leu	UCG	Ser	UAG	Stop	UGG	Trp	A
C	CUU	Leu	CCU	Pro	CAU	His	CGU	Arg	U
	CUC	Leu	CCC	Pro	CAC	His	CGC	Arg	C
	CUA	Leu	CCA	Pro	CAA	Gln	CGA	Arg	G
	CUG	Leu	CCG	Pro	CAG	Gln	CGG	Arg	A
A	AUU	Ile	ACU	Thr	AAU	Asn	AGU	Ser	U
	AUC	Ile	ACC	Thr	AAC	Asn	AGC	Ser	C
	AUA	Ile	ACA	Thr	AAA	Lys	AGA	Arg	G
	AUG	Met	ACG	Thr	AAG	Lys	AGG	Arg	A
G	GUU	Val	GCU	Ala	GAU	Asp	GGU	Gly	U
	GUC	Val	GCC	Ala	GAC	Asp	GGC	Gly	C
	GUA	Val	GCA	Ala	GAA	Glu	GGA	Gly	G
	GUG	Val	GCG	Ala	GAG	Glu	GGG	Gly	A

Figure 1.9A The code by which amino acids for proteins are coded in the nucleic acid RNA. The DNA code is the same except that thymine (T) is substituted for uracil (U). Note the redundancy arising because there are 64 possible 3 letter 'words' coding for only 20 amino acids.

protein in a nearby region turns OFF production of a protein for which the code is in a nearby stretch of the DNA.)

Appendix 1.3 Determining the Age of the Earth

To use the isotopic composition of uranium isotopes in the earth to estimate the age of the earth, one needs to have a way of fixing the initial isotopic composition when the earth was formed. Various methods have been used, but the one which is reported to give the most consistent results is this [45]: Data on the amount of two uranium isotopes and three lead isotopes is assembled for various rocks found in meteorites and the surface of the earth and determined by other methods to be very old (but not dated exactly). It is assumed that all these samples originally had the same isotopic composition of the three lead isotopes but not the same initial composition of the two uranium isotopes when the earth was formed. The isotopes in question are 235-U,decaying to 207-Pb and 238-U decaying to 206-Pb with half lives of $T_{235} = .713$ billion years for 235-U and $T_{238} = 4.509$ billion years for 238-U. 204-Pb is not the endpoint of any such sequence of decays and it remains at the same concentration over very long times. In the following I adopt common practice in the field and use the name of the isotope followed by parentheses indicating the time to denote the amount (number of nuclei) of the isotope present in a sample at that time. Thus $^{235}U(t)$ will mean the number of nuclei of type ^{235}U present in the sample at time t. I will take $t = 0$ to be the time that the earth formed and t to be our present era. The times involved here (hundreds if millions of years to tens of billions of years) are so long that the present does not need to be any more precisely specified. The fact that I refer to a 'sample' and the numbers would be very large and vary from sample to sample will not matter because we will deal with ratios of these numbers for a given sample, which are dimensionless fractions comparable from sample to sample. The definition of half life, then tells us that the quantities defined are related by

$$^{206}Pb(t) =^{206} Pb(0) +^{238} U(0)(1 - (1/2)^{t/T_{238}})$$

because the sample originally had both the lead and the uranium isotope in it but now (at time t) it has more of the lead isotope and less of the uranium isotope in it. The measurable quantity here is $^{206}Pb(t)$. But using the definition again

$$^{238}U(0) =^{238} U(t)(2)^{t/T_{238}}$$

and inserting that back in the preceding equation

$$^{206}Pb(t) =^{206} Pb(0) +^{238} U(t)((2)^{t/T_{238}} - 1)$$

The only unmeasurable quantity here except t itself is $^{206}Pb(0)$. Another such relation can then be written for the second uranium isotope

$$^{207}Pb(t) =^{207} Pb(0) +^{235} U(t)((2)^{t/T_{235}} - 1)$$

Comparison of samples and deduction of the age is made possible by using the fact the third isotope of lead, namely $^{204}Pb(t)$ is not the end point of any decay sequence and can be assumed to be the same at all times in a given sample so that $^{204}Pb(t) = {}^{204}Pb(0)$. Divide both of the previous equations by $^{204}Pb(t)$:

$$^{206}Pb(t)/^{204}Pb(t) =^{206} Pb(0)/^{204}Pb(0) + (^{238}U(t)/^{204}Pb(t))((2)^{t/T_{238}} - 1)$$

$$^{207}Pb(t)/^{204}Pb(t) =^{207} Pb(0)/^{204}Pb(0) + (^{235}U(t)/^{204}Pb(t))((2)^{t/T_{235}} - 1)$$

All the ratios referring to the present (time t) can be measured in all the samples. It is assumed that all the samples formed with the same initial lead composition (but not the same uranium composition). Then the terms $^{206}Pb(0)/^{204}Pb(0)$ and $^{207}Pb(0)/^{204}Pb(0)$ are the same for all samples. Consider a pair of such samples, labeled a and b. To simplify the notation, we denote the ratio $^{206}Pb(t)/^{204}Pb(t)$ by x and the ratio $^{207}Pb(t)/^{204}Pb(t)$ by y. Subtract the x value,x_a, for a from the corresponding value x_b for b using the first of the above equations giving

$$x_a - x_b = ((^{238}U(t)/^{204}Pb(t))_a - (^{238}U(t)/^{204}Pb(t))_b)((2)^{t/T_{238}} - 1)$$

and for y

$$y_a - y_b = ((^{235}U(t)/^{204}Pb(t))_a - (^{235}U(t)/^{204}Pb(t))_b)((2)^{t/T_{235}} - 1)$$

Divide the second equation by the first:

$$\frac{y_a - y_b}{x_a - x_b} = \frac{(((^{235}U(t)/^{204}Pb(t))_a - (^{235}U(t)/^{204}Pb(t))_b)((2)^{t/T_{235}} - 1)}{((^{238}U(t)/^{204}Pb(t))_a - (^{238}U(t)/^{204}Pb(t))_b)((2)^{t/T_{238}} - 1)}$$

Multiply the top and bottom of the right hand side by the ratio $k =^{238} U(t)/^{235}U(t)$ which is assumed and apparently observed, to be the same for all samples giving

$$\frac{y_a - y_b}{x_a - x_b} = \left[\frac{((^{238}U(t)/^{204}Pb(t))_a - (^{238}U(t)/^{204}Pb(t))_b)}{((^{238}U(t)/^{204}Pb(t))_a - (^{238}U(t)/^{204}Pb(t))_b)}\right]\left(\frac{((2)^{t/T_{235}} - 1)}{k((2)^{t/T_{238}} - 1)}\right)$$

The factor in square brackets on the right is just 1 giving

$$\frac{y_a - y_b}{x_a - x_b} = \left(\frac{((2)^{t/T_{235}} - 1)}{k((2)^{t/T_{238}} - 1)}\right)$$

To use this result to get the age, Patterson [45] plotted data on x and y as obtained from meteorites (assumed to be the same age as the age of the earth) against one another as shown in the next figure. They fall on a straight line and a fit to the

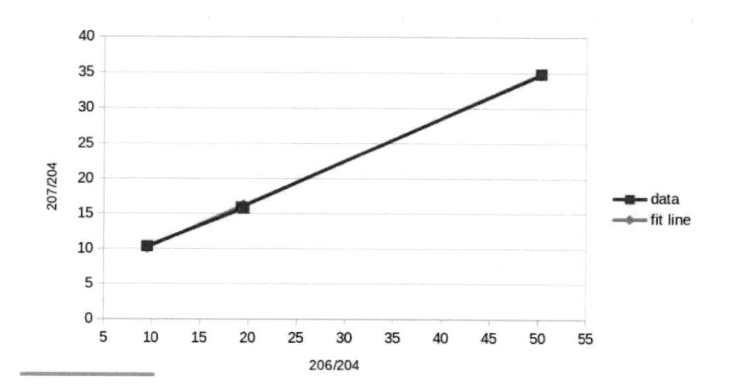

Figure 1.10A Values of the isotope ratios $^{207}Pb/^{204}Pb$ and $^{206}Pb/^{204}Pb$ of some meteorites used in determination of the age of the earth.

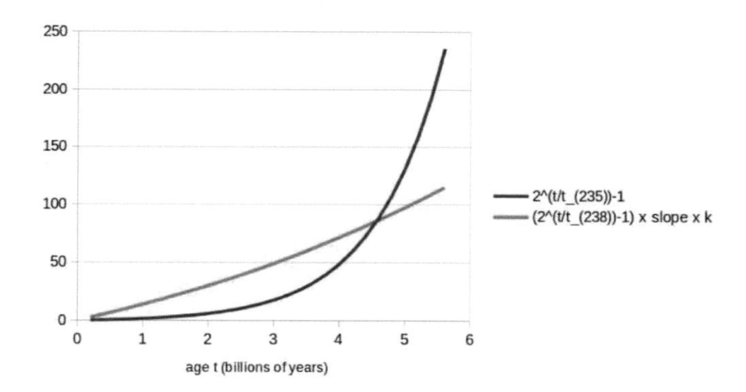

Figure 1.11A Illustration of graphical solution of the equation for the age of the earth. The left hand and the right hand side of the last equation are plotted as a function of the time t . The age estimate is the value of t at which the two curves cross.

formula for a straight line gives a slope of .60256 . Rearranging the last equation to find the age we have

$$(2^{t/T_{235}} - 1) = slope \ \times \ k \times (2^{t/T_{238}} - 1)$$

The observed value of $k =^{238} U(t)/^{235}U(t)$ (referring to the present) is about 138.7. Thus we can plot the right hand side versus the lefthand side of this equation versus t and the inferred value of the age of the earth is where the curves cross. The graph is shown in Figure 1.11A and gives an age of about 4.5 billion years as you can see.

Appendix 1.4 Doppler Shifts

The determination of the speed at which distance galaxies are receding from earth depends almost entirely on the phenomenon of a doppler shift in the light emitted by a receding distance object. Here I describe the phenomenon as it occurs for relatively nearby stars where effects of general relativity are not very important and I will neglect them. The first point to understand is that, in laboratory experiments on earth, a vapor consisting of any atom of the periodic table will, if electrically excited, emit lines of characteristic colors which can, among other uses, be employed to identify the atoms in the gas. (The emission is familiar in old style fluorescent lights.) This feature is used regularly to discover the atomic content of stars by studying the colors (more precisely the spectra) of the light which they emit. Colors are associated with wavelengths of the emitted electromagnetic radiation (Appendix 1.1) so we may envision light of a characteristic wavelength emerging from a star and traveling a vast distance into the telescope of an astronomer (or these days, of a vast team of astronomers) as sketched in the cartoon in Figure 1.12A.

In the first panel, we sketch a wave form for a light wave emitted by a star which is moving at a negligibly small speed with respect to the earth. By measuring the wavelength λ of the light when it arrives in the telescope, astronomers can identify the light as characteristic of an atom in the star and the wavelength will be the same as that characteristic of the atom when it is excited electrically in the laboratory or in a fluorescent light on earth. By contrast consider, as sketched in the second panel, what happens when the star is moving way from earth at a significantly high speed v. The atoms in the star are still emitting peaks in the wave at the same frequency, ie with the same time T, the period, between peaks as they would be if the star were not moving away. But after one peak is emitted, the star moves away a distance $v \times T$ before it emits the next peak. Therefore the distance between the peaks of the light which the astronomer observes in the arriving light is $\lambda + T \times v$. The light is 'stretched' to longer wavelengths. Since longer wavelengths correspond to redder light this shift in the wavelength is called a 'red shift' as well as the 'Doppler shift' after the person who first explained it. Because the wavelength, the velocity of the wave, called c, and the period of the wave are related by $\lambda = cT$ we have a red shift which can be written as $\Delta\lambda = \lambda(v/c)$. Astronomers denote the ratio $\Delta\lambda/\lambda$ by z so in this case of small but not negligible red shifts, we have $z = v/c$.

As discussed in Chapter 1, the Hubble constant is the ratio v/d where d is the distance between earth and the observed object. The distance is much the more difficult quantity to measure. One way of doing it is described in the next appendix where I will illustrate the determination of an estimate of the age of the universe using some data from nearby galaxies.

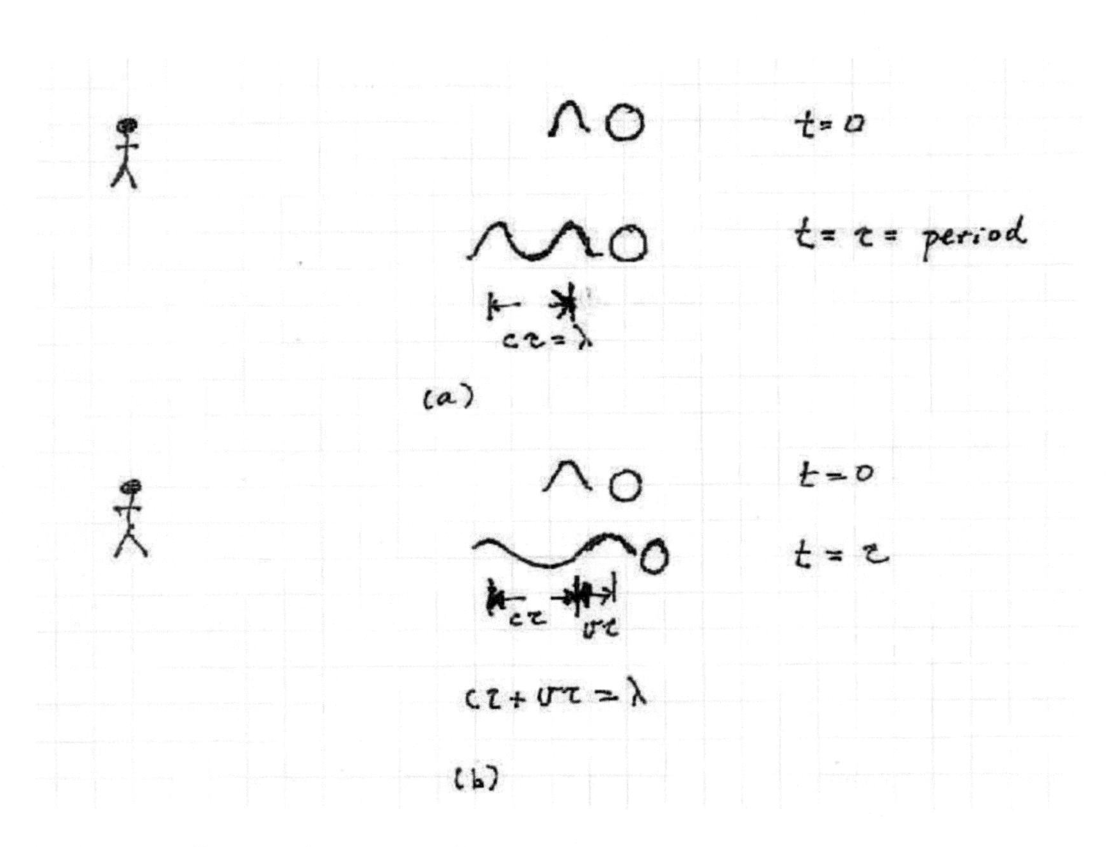

Figure 1.12A Cartoon illustration of Doppler shift.

Appendix 1.5 Determination of Distances of Galaxies from Earth and Estimates of the Age of the Universe

Only the method using supernovae as 'standard candles' will be described here. Another method, using data on the cosmic microwave background which is a kind of 'echo' of the big bang involves adjusting equations from a theory of how the bang occurred to the microwave data. The fitting involves seven parameters, one of which is the age of the universe. That method is less direct than the one described here, it is dependent on the correctness of the theoretical model and its meaning is more difficult to grasp intuitively. Nevertheless it gives results which are quantitatively similar those found using the method described here.

As described in Chapter 1, an estimate of the age of the universe can be obtained from the inverse of the Hubble constant which is defined as $H = v/d$ where v is the velocity with which a galaxy is receding from earth and d is its distance from earth at the present. To illustrate how this works, consider Figure 1.10 which is a graph of the distance of two galaxies 1 and 2 from the earth as a function of time t. $t = 0$ is taken to be the present and since the galaxies are receding, the slope of the graphs will be positive corresponding to the fact that they were closer to us in the past. Astronomers measure the quantities $r_1(0), r_2(0), v_1$ and v_2, all in the present. How the distances are measured is described in this appendix and the velocity measurements are made by measuring the Doppler shift as described in the preceding appendix. If the graphs of distance versus time are straight lines, as assumed in Figure 1.10 then those numbers are enough to permit the lines to be drawn on the graph using the relations $r_1(t) = r_1(0) + v_1 t$ and $r_2(t) = r_2(0) + v_2 t$. (All the numbers on the right hand side have been measured except for t.). At the big bang, the two galaxies were presumed to have been in the same place, so the distances from earth would be zero. Setting both $r_1(t_1) = 0$ and $r_2(t_2) = 0$ in the above equations gives $t_1 = -r_1(0)/v_1$ for the time when galaxy 1 was on top of us and $t_2 = -r_2(0)/v_2$ when galaxy 2 was on top of us. The remarkable fact is that, using the observational data, those ratios of d/v come out almost the same for almost all galaxies. That corresponds to the straight lines in figure intersecting at $r = 0$ and the same $t = -d/v$ as shown in the figure and, physically, to the inference that the galaxies were on top of us at the same time $-d/v = -1/H$.

It is not at all obvious that the values of d/v should be the same for all receding galaxies but Hubble suggested and it has been since shown from more precise and extensive observations to be so to a good approximation. The estimate of $1/H$ as the age of the universe assumes that the recession velocities have remained the same throughout their history and that is certainly not the case. It would be exactly correct if the universe were massless, but of course that involves a contradiction since in that case there would be no distant galaxies to measure. Figure 1.11 shows how the distance between galaxies changes versus the time in various theoretical models of the history of the universe. It would give a straight line if the velocities remained constant. Approximating those curves by a straight line with a slope equal to its slope at the present would give the age of the universe as the inverse of the Hubble constant. From Figure 1.11 one can see that though a straight line is not a perfect match to those curves, it can be expected to give a reasonable order of magnitude estimate for the age. Confining our goal to making such an estimate it remains to understand how to estimate the Hubble constant from observational data on receding galaxies.

Recalling that $H = v/d$ the measurement of the red shift in the spectra of light from the stars in the galaxies as described in Appendix 1.4 gives quite an accurate estimate of the recessional velocity v and is the easier part of the observational problem. The much harder part is determining the distance d of the galaxies from earth at the present time. That problem has plagued astronomers for centuries. In the present context it is the reason that Type 1a supernovae were chosen for recent estimates. Supernovae are exploding stars. All stars shine because the gravitationally compressed materials in their interiors undergo nuclear fusion reactions which release large amounts of energy which in turn excites the electrons attached to atoms nearer to their surfaces which finally emit the light which we observe. This process of 'nuclear burning' can go on for billions of years if the star is not too massive because the gravitational pressure is not as extreme as it is in heavier stars (For our star, the sun, the estimated lifetime is of the order of 10 billion years.) For heavier stars, the burning occurs faster (more gravitational pressure) and the nuclear fuel can quite suddenly run out, causing a collapse and a bounce manifested as a flash of light and other types of radiation in the sky. Supernovae have been observed by astronmers, rather infrequently until recently, for hundreds of years. The light rapidly grows less intense after the initial flash but remanents of supernovae that occurred hundreds of years ago have been observed. In the context of estimates of the Hubble constant the relevance of supernovae is that, using improved technological methods, astronomers are now able to detect thousands of supernovae in distant galaxies in just a few years. Among them a certain type of supernova, called Type 1a, was found to almost always emit light of the same brightness (technically intensity). That permitted a distance estimate (the factor d in the definition of the Hubble constant) as follows: Once the light leaves the supernova it experiences very little interaction with other matter in the vacuum of space so in first approximation we can assume that it carries the same energy that it had when it left the supernova when it arrives

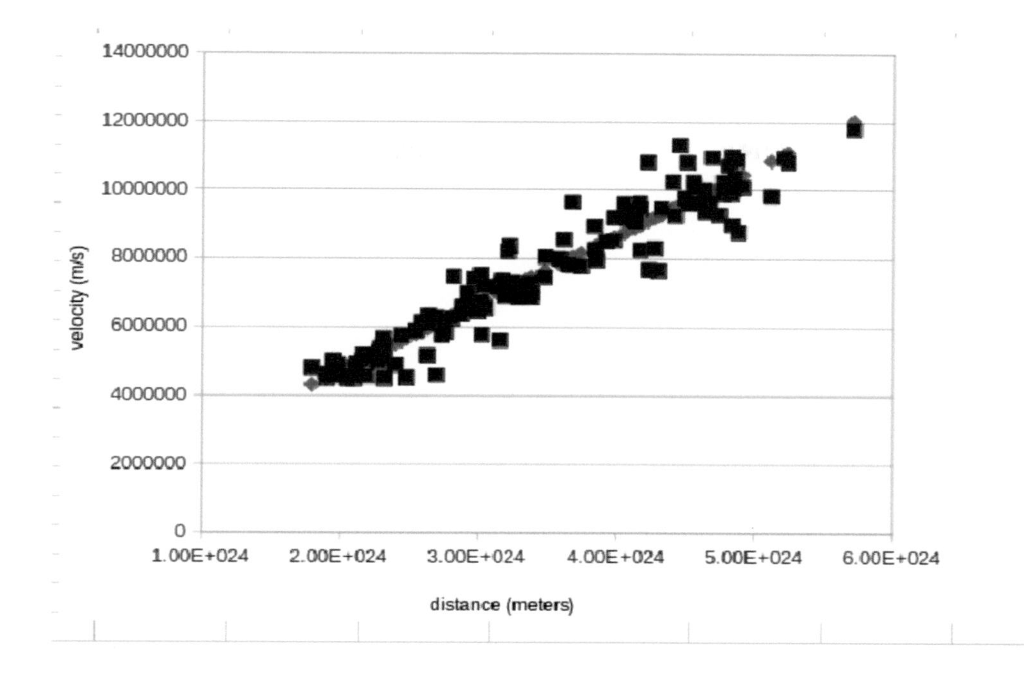

Figure 1.13A Velocity versus distance for data on nearby galaxies. The slope as determined from fitting a straight line to this data gave an estimate of the Hubble constant corresponding in the case that the age of the universe estimated as 1/(Hubble Constant) by about 16.1 billion years.

at the astronomers' telescopes. But the explosions are nearly spherically symmetric, that is the energy flies off in all directions, so it gets spread over a sphere of radius d when it gets to earth. Denoting the energy of the light from a supernova by E (the same for all type 1a supernovae), the light coming into the astronomer's telescope has energy of about $A \times E/4\pi d^2$ where A is the area of the telescope opening which is receiving the light. Lower energy means that the supernovae is perceived as dimmer, so more distant supernovae are dimmer in a way which permits d to be estimated for each one by measuring the energy in the light observed. That is basically the way that the distances in publicly accessible data on 126 nearby galaxies which are openly available in [33] were determined. A graph of the velocity as a function of the distance as given in that data appears in Figure 1.13A.

One can get an idea of the kind of analysis required to get the age of the universe as shown in that figure by fitting the data to a straight line whose slope will be the Hubble constant if the Hubble relation is obeyed. The Hubble relation is approximately obeyed in this data set but the value of the velocity at zero distance is a small finite number for which the Hubble relation gives zero. From the value of the Hubble constant one then infers the time which has passed since the big bang (the 'age' of the universe) using a model of the way the expansion occurred after the bang. If the universe were massless that would give an age of 1/(Hubble constant) which, for this data set, is about 16.1 billion years, higher than inferred from more complete data and better models, but in the right ball park.

Appendix 2.1 Defining the Instantaneous Present and Predicting the Future with Newtonian Physics

Here are some details indicating how a laptop computer can be used to make predictions of the future using Newtonian physics. A laptop will only work with rather simple examples like the ones discussed here but essentially identical methods are used by professionals to predict the behavior of millions of interacting particles, including those found in living systems.

Within Newtonian physics, everything is described by the positions of material particles at each instant in time. That doesn't present any serious conceptual problems, though atoms were unknown in Newton's day so the identity of the particles was not specified. What was conceptually more challenging was that, at each instant of time (eg at the present) a complete description of the Newtonian world also required the specification of the velocities of all the particles. If you think of speed as the magnitude of the velocity and define speed as

speed = (distance gone)/(time elapsed)

then there is a puzzle because the present is thought to be infinitesimally short in time so no time can have elapsed during it.

Newton resolved the issue by defining a quantity called instantaneous velocity, which we will discuss, However to illustrate the nature of the problem further and illustrate that it had been troubling for a long time let's briefly consider one of Zeno's paradoxes: Zeno was a philosopher who lived in Italy about 2500 years ago. One of his paradoxes concerned an arrow in flight. In modern terms, it is easiest to describe in terms of photography, though of course Zeno had no cameras. If you look at a still photograph of the arrow, which presumably is representative of the state of the arrow at the instant that the photo was taken, you generally CANNOT TELL IF THE ARROW IS MOVING OR NOT. (For example, it could be hanging by a very fine thread not visible in the picture. In fact that is a common way to fake pictures of flying saucers.)

The picture is only telling us the position of the arrow, and the example illustrates why Newton could not be content to describe the state of a system in his theory by only specifying positions of things at each instant.

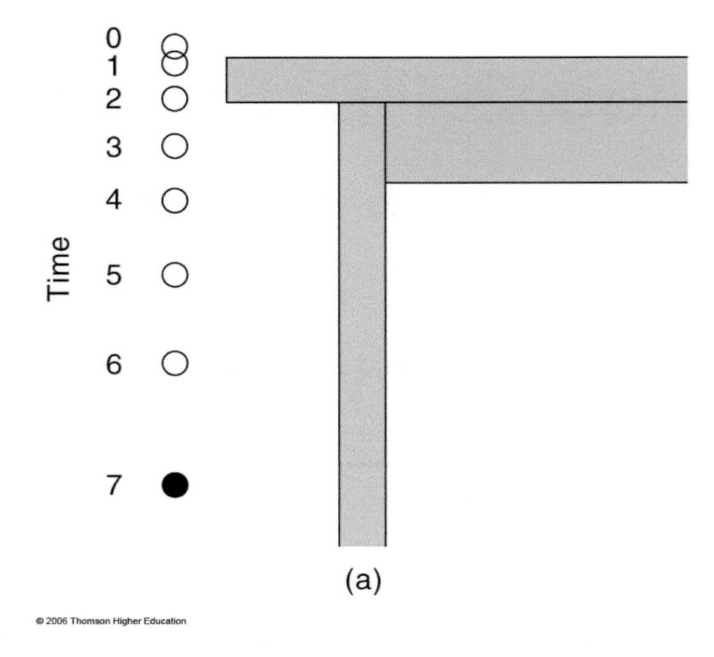

Figure 2.1A Equal time shapshots of a falling ball.

To understand the way out, think about how you would find out whether the arrow in the picture is moving. Ideally, the picture might be one still (or frame) from a movie. Then you could look at a few frames just before the instant in question. If the arrow is in different positions in those neighboring frames you are confident that it is moving at the instant of the original picture. But those other frames are characteristic of the past, not of the present. How do we get quantities which only involve the present and which tell us if the arrow (or anything else) is moving? Well, you (and Newton) do it like this. Look at a bunch of frames in the past of the arrow picture, each one a little bit closer to the present than the previous one. For each frame you can determine the position of the arrow at that previous instant and use the equation:

speed = (distance gone)/(time elapsed)

to estimate the speed at the present instant. As the frames get closer and closer to the present one, the time elapsed gets smaller and smaller but so does the distance gone. You can make a table of the speeds and the elapsed times and plot up the results. An example of the result of such a procedure appears in Figure 2.3. You can then use such data to infer a speed for zero time elapsed by extending the curve obtained by connecting the dots to zero time elapsed.

Here is another example, referring to a falling object: Using a fast camera you can get a series of pictures of an object, such as a ball, dropped straight down. (It's a common exercise in physics courses.) Such a series of pictures is shown in Figure 2.1A and the corresponding data is shown in Figure 2.2A.

Time (seconds)	Distance Fallen (meters)
0	0
0.033	0.017
0.067	0
0.1	0.034
0.133	0.06
0.167	0.102
0.2	0.144
0.233	0.203
0.267	0.254
0.3	0.355
0.333	0.457
0.367	0.508

Figure 2.2A Data on a falling ball.

To get a number for the instantaneous velocity 0.3 seconds after it was dropped you can use this data to obtain the following table in Figure 2.3A from the definition of speed. (Here the motion is all in a straight line down so the velocity is the speed straight down.) And a graph of this as shown in Figure 2.4A shows how the extrapolation to zero time elapsed is done.

Thus you can use data from the past and to get a quantity, the instantaneous velocity, which is characteristic of the present instant. Those instantaneous velocities

x(m)	times	interval	velocity estimate
0			
0.017	0	0.3	1.18333333333333
0	0.033	0.267	1.26591760299625
0.034	0.067	0.233	1.52360515021459
0.06	0.1	0.2	1.605
0.102	0.133	0.167	1.76646706586826
0.144	0.167	0.133	1.90225563909774
0.203	0.2	0.1	2.11
0.254	0.233	0.067	2.26865671641791
0.355	0.267	0.033	3.06060606060606
0.457	0.3	0	
0.508			

Figure 2.3A Estimates of the velocity of the dropping ball.

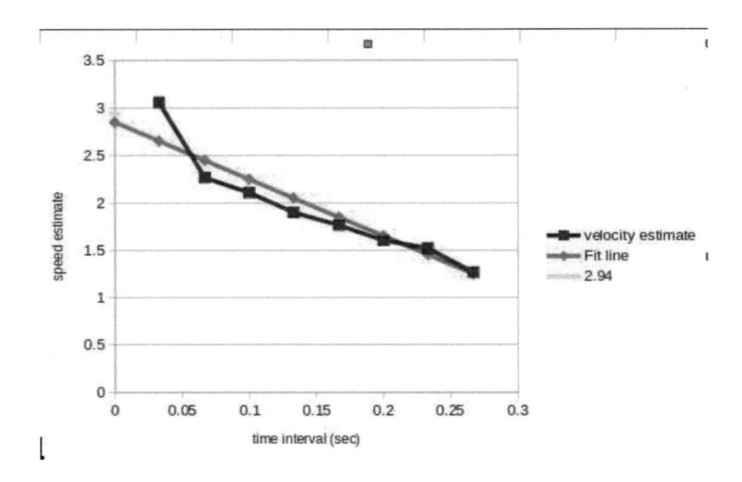

Figure 2.4A Extrapolation of the data to get the instantaneous velocity.

are absolutely central to making Newton's theory work. He must characterize each instantaneous moment in time by positions AND instantaneous velocities. You can see that the latter implicitly contain information about the past through the limiting process I have described. So in a sense, one can say that Newton regards the present as an infinitesimally short instant, but the description of that instant depends on use of information about the immediate past (or the future but of course we don't have information about the future at the present moment.)

So far, I've described how Newton conceived and characterized the present: It is an infinitesimally thin slice of time characterized by the positions and instantaneous velocities of all the particles (in general) in the universe. I emphasized that determining instantaneous velocities involved using some information about the past.

Now what about the future? Newton found that at least in many cases, he could PREDICT the future of at least part of the universe given the information he required to characterize the present. This could be done if he had a correct mathematical model of the forces on all the particles in the present and he knew their masses.

So what's a Newtonian force?

It is related to, but not necessarily the same as, one's intuitive idea of force. To understand it, first one has to understand the idea of instantaneous ACCELERATION in the present instant. It is related to the instantaneous velocity in the same way that instantaneous velocity is related to position, by a limiting process. Call the instantaneous acceleration of some particle at the present time t a(t). Then the formula that makes prediction possible is this equation which can be regarded as Newton's DEFINITION of the force on a particle.

$$\mathrm{Ma(t)} = \mathrm{F(t)}$$

where M is the mass of the particle and F(t) is the force on the particle at the present instant. You have to have, in addition, a THEORY which tells you what the force is, given the instantaneous positions. Newton's theory of planetary motion used his universal law of gravitation for determining the needed forces. That theory is still useful if the forces of gravitation are not too large. However the power of Newton's method is that other forces, such as electrical and magnetic ones, can also be included within the same framework to make predictions if those forces are present.

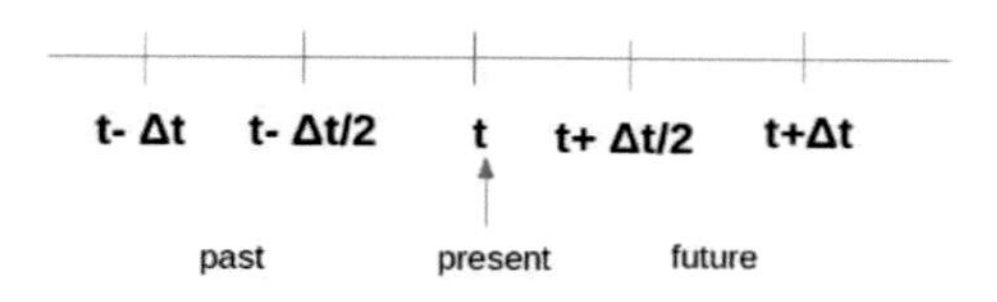

Figure 2.5A Times associated with Newtonian prediction.

Now confining attention to just one particle moving in just one dimension, I will show how the future can be predicted as precisely as you like if you have some information about the recent past and a model (usually an equation) which gives the force at the present in terms of the positions at the present. I will need to define a few symbols: We let t be the present time (a number in seconds, for example) and Δt a small interval of time. In a precise calculation using Newton's theory we would need to find results for the case that Δt shrinks to zero as in the discussion of instantaneous velocity above. However it turns out that by judicious choice of a small enough, but finite, value of Δt one can get quite precise predictions. I consider 5 times: $t - \Delta t$, $t - \Delta t/2$,t,$t + \Delta t/2$ and $t + \Delta t$. The first two are in the past, the middle one (t) is the present and the last two are in the future. They are laid out in a line of times as in Figure 2.5A.

As noted, Newton thought of time as a quantity with values on such a line of real numbers stretching infinitely far into the past and the future. We use the symbol x for the position of the particle, v for its velocity and a for its acceleration. All three quantities can change as time progresses, as indicated by a time argument. For example, the velocity at time $t-\Delta t/2$ is denoted $v(t-\Delta t/2)$. It is given approximately as

$$v(t - \Delta t/2) = (x(t) - x(t - \Delta t))/\Delta t$$

referring to the velocity in the recent past and

$$v(t + \Delta t/2) = (x(t + \Delta t) - x(t))/\Delta t$$

referring to the velocity in the near future. The acceleration at the present is defined in terms of these:

$$a(t) = (v(t + \Delta t/2) - v(t - \Delta t/2))/\Delta t$$

Recall that we are assuming that we know how to calculate the force $F(t)$ in terms of properties of the present and that $F(t) = Ma(t)$. Now it is not hard to rearrange these equations, putting the things that refer to the future on the left and the things that refer to the past and the present on the right and eliminating a(t) in favor of the force F which we suppose that we know because it is defined in terms of present positions. The result of the rearrangement is

$$\begin{array}{rcl} \textit{future} & & \textit{past} \qquad\qquad \textit{present} \\ x(t + \Delta t) & = & v(t - \Delta t/2)\Delta t + x(t) + (F/M)\Delta t^2 \\ v(t + \Delta t/2) & = & v(t - \Delta t/2) \qquad + (F/M)\Delta t \end{array} \tag{2.1}$$

These formulas then tell you how to predict the immediate future from data about the past and the present. Admittedly, it is a very near future, but by using this formulation repeatedly you can inch forward from the present to the future indefinitely. By writing essentially identical relations for up to billions of particles, scientists predict the future of many complex systems. The accuracy of the predictions depends on several considerations, some of which are discussed in Chapter 3. Clearly you must have an accurate theory of the forces. Also, since Δt is finite there can be errors because the theory really only works in the limit that Δt is zero. However those two constraints are under good control in professional codes. A more challenging problem, both conceptually and practically, is that, when there are many particles interacting through the forces, many, in fact probably most, Newtonian systems representing the real world give futures which are extremely sensitive to the starting postions and velocities (the 'initial conditions'). That feature, discussed further in Chapter 3, is related to the idea of chaos and plays a role in the physics of heat. Even with three particles, many Newtonian systems have that property and it can be seen in the system for which simulation results were shown in Figure 2.2

It is quite easy to use the last two equations to make predictions about simple systems like falling balls and pendula using an Excel spreadsheet. Ambitious readers wanting to gain a detailed feeling for the meaning of the formulation are encouraged to try it.

Appendix 3.1 Coarse Graining in Card Games

Somewhat as was done in Chapter 3 with dice, here we consider another 'toy universe' consisting of a deck of cards. Players are dealt a 'hand' of 6 cards from a standard deck in which there are 52 cards in 4 suits, with each suit having a number from 1 to 13. Two of the suits are red and two of them are black. (This is all standard if you have played cards.)

Now consider the following specifications of the state of your hand of cards:

a) you specify the number and suit of each card and specify the order in which it was dealt to you.

b) you specify the number and suit of each card but not the order in which it was dealt to you.

c) you specify only the color of all the cards. Say there are r red ones and 6-r black ones. r can take values from 0 to 6.

The 'microscopic' specification of the state is case a) and the most 'coarse-grained' one is case c. As an exercise, you can work out the entropy of each, using the definition of entropy given in Chapter 3. Notice that, in case c, the answer depends on r, that is, at that coarse-grained level of description, some states have higher entropy (and are more likely to be dealt to you) than others. The state of highest entropy ($r = 3$) is the most likely one. So if you started with low entropy state, say $r = 0$, and exchanged cards repeatedly with those in the remaining deck, it would be likely, but not certain, that you would end up with a hand in the $r = 3$ state. On average, the entropy would increase, though there would be improbable fluctuations from time to time in which it would go down.

All these features illustrate similar ones in a more complex universe of billions of atoms interacting over time with forces that can be described in a Newtonian manner. Thus the toy model illustrates how the second law manifests itself in the everyday world we experience.

Detailed answers for the entropies in the three cases are given on the next page.

One possibility. $S/k = ln(1) = 0$

6 ways to get the first $\times$ 5 ways to get the second ..$=6 \times 5 \times 4 \times 3 \times 2 \times 1 = 6!$ $S/k = ln(6!) = 6.579$

r = 1:

5 blacks, 1 red. 26 ways to get 1st black $\times$ 25 ways to get second ... $\times$ 26 ways to get the red times (6!/5!1!) ways to mix the reds and blacks.

$S/k = ln((26!/21!)(26!/25!)(6!/5!1!)) = 20.93$

r = 2: : 4 blacks, 2 reds: $S/k = ln(26!/22!)(26!/24!)(6!/(4!2!)) = 21.97$

r = 3:

3 blacks, 3 reds: $S/k = ln((26!/23!)(26!/23!)(6!/(3!3!))) = 22.30$

r = 4:

S/k = 21.97

r = 5:

S/k = 20.93

r = 6

S/k = 18.92

Appendix 5.1 Michelson Morley experiment

The experiment was intended to find the velocity of the earth relative to a presumed stationary medium, called the 'ether' in which light waves propagate. The reasoning was identical to that applied to analyze the water wave experiment described at the beginning of Chapter 5, with the earth playing the role of the boat and the 'ether' of the water. The setup consisted of a light source, three mirrors and a detector (see Figure 5.1A). One of the mirrors was 'half-silvered' so that it let half the light incident on it through and reflected the other half. That half-silvered mirror was placed near a light source with its plane at 45 degrees to the direction of the incoming light so that it reflected half the light from the source along the direction of the presumed velocity of the earth relative to the ether and the rest of the light passed through in a direction perpendicular to the direction of the velocity of the earth. The two other mirrors (totally reflecting) were placed in the paths of those perpendicularly oriented beams of light and reflected them back to the half-silvered mirror where the beam moving back against the motion of the earth passed through the half-silvered mirror and the beam coming back in a direction perpendicular to the direction of the earth's motion was reflected (downward in the figure). As a result the two beams were combined at the detector and the light waves could interfere.

Now Michelson and Morley figured, following the reasoning of the boat problem, that the upward moving beam would be moving a speed $c - v$ in the earth bound laboratory and that on its return trip it would be moving with speed $c+v$. Calling the distance between the half-silvered mirror and the upper mirror L they thus concluded that the time for the round trip of a pulse of the light from the source would be

$$\Delta t_{||} = L(1/(c-v) + 1/(c+v)) = (2L/c)/(1-(v/c)^2)$$

(Despite some similarity in the formulas, this calculation took no account of relativity and is, to be clear, physically wrong.)

The round trip time for the light moving perpendicularly to the direction of the earths presumed velocity was calculated by noting that the mirror to the right in the figure was moving vertically that is along direction of the earth's motion, during the trip. Assuming that that mirror was also a distance L from the half-silvered mirror one can see that the light would have to travel a distance $\sqrt{(L^2 + (v\Delta t_\perp/2)^2)}$ using the Pythagorean theorem as shown in the bottom part of the figure. To find the time $\Delta t_\perp$ for that round trip one thus has to solve the equation:

$$\Delta t_\perp/2 = \sqrt{(L^2 + (v\Delta t_\perp/2)^2)}/c$$

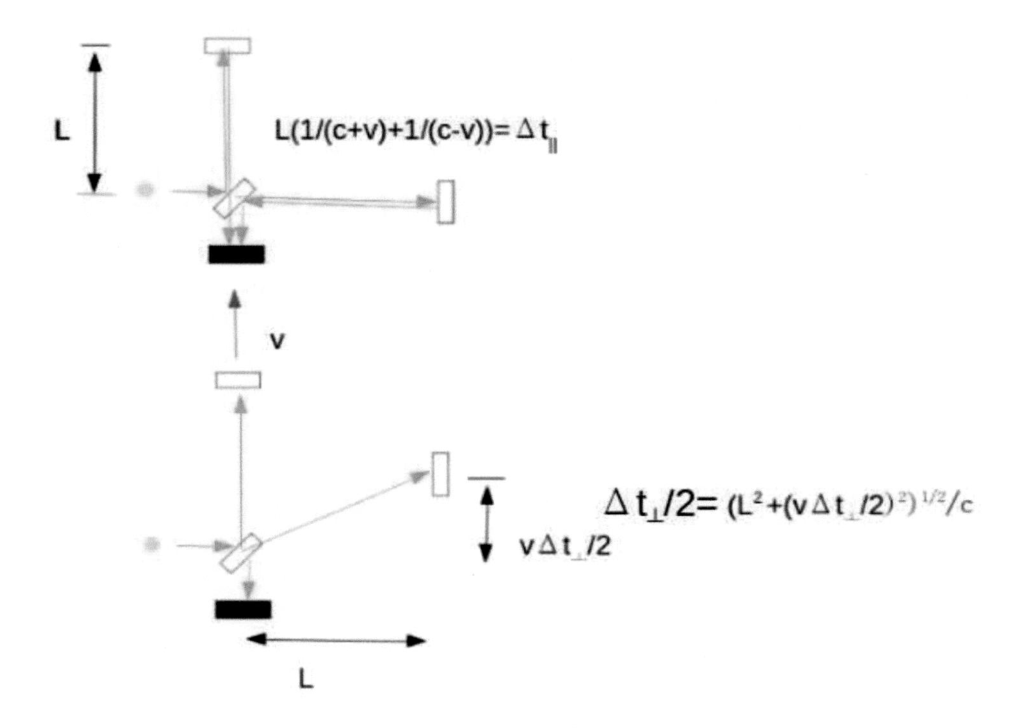

Figure 5.1A Essential features of the Michelson Morley apparatus.

for $\Delta t_\perp$ That is not hard and gives

$$\Delta t_\perp = (2L/c)/\sqrt{1 - v^2/c^2}$$

Thus, according to this calculation $\Delta t_{||} \neq \Delta t_\perp$ so that wave peaks in the two waves would arrive at different times resulting in interference and reduced intensity when they were mixed together and detected. Michelson and Morley did observe the dark lines in the intensity of light characteristic of such interference. To see if part the interference pattern was affected by the motion of the earth through the ether, they rotated the apparatus through 90 degrees so that the path previously along the direction of the earth's motion was perpendicular to it, and the path previously perpendicular to the direction of the earth's motion was along it. The observed pattern of interference was then expected to shift, whether the distances were exactly the same or not. But they saw no shift. The ultimate conclusion was that the time for the light to go up, down and sideways in the apparatus was exactly the same (L/c), because even if the earth was moving through a medium, as they had postulated, the velocity of light was the same in the moving earth as it would have been in a postulated stationary frame. That was consistent with Einstein's postulate, but it took physicists some time to accept because it is very nonintuitive.

Appendix 5.2 The 'G' in the Lorentz Transformation

We show here that, if as required by Einstein, the same formulas will work to describe the space and time separations between pairs of events in both directions, from one frame to another and back, then the square of the prefactor G in those relations is a unique function of the speed of one of the frames with respect to the other.

As described in Chapter 5, the relations in one direction, giving the time and space separations $\Delta t', \Delta x'$ in a frame moving with velocity v with respect to another in which the separations are $\Delta t, \Delta x$ are

$$\Delta t' = G(\Delta t - (v/c^2)\Delta x)$$

$$\Delta x' = G(\Delta x - v\Delta t)$$

The form of the second relation is the same as that of the Galilean transformation except for the factor G and the first relation, for $\Delta t'$, was shown to need the second term involving Δx in order that the speed of light come out the same in all frames, as experiment required.

The transformation in the other direction (from the primed to the unprimed frame) then must have the form

$$\Delta t = G(\Delta t' + (v/c^2)\Delta x')$$

$$\Delta x = G(\Delta x' + v\Delta t')$$

because from the primed frame the unprimed frame appears to be moving 'backwards', that with velocity $-v$. To find out how to make the two sets of relations compatible, we take the expressions in the first pair of equations and substitute the right hand sides into the first of the second pair of equations:

$$\Delta t = G(G(\Delta t - (v/c^2)\Delta x) + (v/c^2)G(\Delta x - v\Delta t))$$

Take a factor G out of each term on the right giving

$$\Delta t = G^2(\Delta t - (v/c^2)\Delta x + (v/c^2)\Delta x - (v/c)^2\Delta t)$$

The two terms involving Δx cancel leaving

$$\Delta t = G^2\Delta t(1 - (v/c)^2)$$

Divide out Δt and divide both sides by $(1 - (v/c)^2)$ giving

$$G^2 = 1/(1 - (v/c)^2)$$

as claimed.

You can show that the same relation arises if you start with the equation for Δx and make the corresponding substitutions.

Appendix 5.3 Proper Time Intervals are the Same in All Frames

Start with the definition of the square Δs^2 of the proper time interval between a pair of events as expressed in an unprimed coordinate system:

$$\Delta s^2 = \Delta t^2 - (\Delta x/c)^2$$

and the Lorentz transformation to another reference frame

$$\Delta t = G(\Delta t' - (v/c^2)\Delta x')$$

$$\Delta x = G(\Delta x' - v\Delta t')$$

Insert those two relations into the definition:

$$\Delta s^2 = (G(\Delta t' - (v/c^2)\Delta x'))^2 - (G(\Delta x' - v\Delta t')^2/c^2$$

Write out the squares and take out a factor of G^2:

$$\Delta s^2 = G^2 \Big[\Delta t'^2 - 2\Delta t'(v/c^2)\Delta x' + ((v/c^2))^2(\Delta x')^2 \\ -(\Delta x'^2/c^2 - 2(\Delta x' v\Delta t')/c^2 + (v\Delta t')^2/c^2)\Big]$$

The cross terms cancel out leaving

$$\Delta s^2 = G^2(1 - (v/c)^2)(\Delta t'^2 - \Delta x'^2/c^2)$$

But (Appendix 5.2) $G^2 = 1/(1 - (v/c)^2)$ so

$$\Delta s^2 = \Delta t'^2 - \Delta x'^2/c^2$$

has the same value when computed using the space and time intervals measured in the primed frame.

Bibliography

[1] A. Abragam. *The Principles of Nuclear Magnetism.* Oxford University Press, London, 1961.

[2] J. R. Ackermann. website:http://www.leapsecond.com/java/gpsclock.htm.

[3] P. W. Anderson. *Science*, 177:39, 1972.

[4] A. Baker. *Simplicity.* Metaphysics Research Lab, Stanford University, winter 2016 edition, 2016.

[5] D. Bohm. *Physical Review*, 85:166, 1952.

[6] N. Bostrum. *Technology Review*, May/June:72, 2008.

[7] H. J. F. Victoria, E. A. Brunsdon, and E. E. F. Bradford. *Scientific Reports*, 11:1382, 2021.

[8] C. Dewdney, C. Philippidis, and B. J. Hiley. *Nuovo Cimento*, 52B:15–28, 1979.

[9] K. Camilleri. *Studies in History and Philosophy of Science Part B: Studies in History and Philosophy of Modern Physics*, 40:290–302, 2009.

[10] N. Carr. *The Shallows What the Internet Is Doing to Our Brains.* W. W. Norton, NY, 2010.

[11] S. Carroll. *From Eternity to Here.* Dutton, NY, 2010.

[12] N. Clauer. *Chemical Geology*, 354:163, 2013.

[13] W. Yu, H. Zheng, J. H. Houl, B. Dauwalder and P. Hardin. *Genes Dev*, 20:723–733, 2006.

[14] P. C. Davies. In *Physical Origins of Time Assymmetry.* JJ Halliwell, J. Perez-Mercader and W. H. Zurek, eds., pages 119–130. Cambridge University Press, 1994.

[15] P. C. W. Davies. *The Physics of Time Assymetry.* University of California Press, Berkeley, CA, 1977.

[16] N. M. Dotson and M. M. Yartsev. *Science*, 373:242, 2021.

[17] J. Klokocnjk et al. Correlation between the mayan calendar and astronomy helps to answer why the most popular correlation (gmt) is wrong. *Astronomische Nachrichten*, 329:329, 2008.

[18] M. Kouduka et al. *International Journal of Plant Genomics*, 2007:Article ID 27894, 2007.

[19] S. A. Benner et al. *Science*, 296:864–886, 2002.

[20] V. Ho et al. *Science*, 334:623, 2012.

[21] D. H. Frisch and J. H. Smith. *American Journal of Physics*, 31:342, 1963.

[22] M. Gellman and J. Hartle. In *Physical Origins of Time Assymmetry.* JJ Halliwell, J. Perez-Mercader and W. H. Zurek, eds., pages 311–337. Cambridge University Press, 1994.

[23] J. Gleick. *Chaos: Making a New Science.* Penguin, New York, NY, 1987.

[24] S. Goldstein. Bohmian mechanics. In *The Stanford Encyclopedia of Philosophy.* Fall 2021 Edition, Edward N. Zalta ed. https://plato.stanford.edu/archives/fall2021/entries/qm-bohm, 2021.

[25] C. Graney. Anatomy of a fall and the story of g. *Physics Today*, pages 36–40, 2012.

[26] J. C. Greene. *The Death of Adam; Evolution and Its Impact on Western Thought.* Iowa State University Press, Ames, 1959.

[27] D. O. Hebb. *The Organization of Behavior: A Neuropsychological Theory.* Wiley, NY, 1949.

[28] P. R. Holland. *The Quantum Theory of Motion.* Cambridge University Press, Cambridge, 1993.

[29] T. Jarrett. *Publications of the Astronomical Society of Australia*, 21:396–403, 2004.

[30] S. A. Josselyn and S. Tonegawa. *Science*, 367:39, 2020.

[31] D. Kahneman. *Thinking, Fast and Slow.* Farrar, Strauss and Giroux, NY, 2011.

[32] I. E. Tamm L. I. Mandelshtam. *J. Phys. (USSR)*, 9, 1945.

[33] Brookhaven National Laboratory. website:http://supernova.lbl.gov/union.

[34] Brookhaven National Laboratory. website:https://www.nndc.bnl.gov.

[35] R. B. Laughlin. *Neuron*, 83:1253, 2014.

[36] S. Laureys and G. Tononi. *The Neurology of Consciousness Cognitive Neuroscience and Neuropathology.* Academic Press, London, 2009.

[37] W. F. Libby. *Science*, 133:621–629, 1961.

[38] L. Maccione. website:https://pirsa.org/21060100, 2021.

[39] S. Cintoli, M. C. Cenni, B. Pinto, S. Morea, A. Sale, L. Maffei, and N. Berardi. *Neural Plasticity*, 2018:3725087, 2018.

[40] J. Minxha, R. Adolphs, S. Fusi, A. N. Mametak, and U. Rutishauser. *Science*, 368:eaba3313, 2020.

[41] T. Nakagawa. *Science*, 366:1259–1263, 2019.

[42] H. Nikolic. *Foundations of Physics Letters*, 19:259, 2006.

[43] J. J. O'Connor and E. F. Robertson. https://mathshistory.st-andrews.ac.uk/HistTopics/Longitude1. *MacTutor History of Mathematics*, 1997.

[44] D. N. Page. Information loss in black holes and/or in conscious beings? In *Heat Kernel Techniques and Quantum Gravity*, page 461. arXiv:hep-th/9411193, 1994.

[45] C. Patterson. *Geochimica et Cosmochimica Acta*, 10:230, 1956.

[46] W. Pauli. *Theory of Relativity.* Pergamon Press Ltd, Oxford, 1958.

[47] H. Poincare. *Les Methodes Nouvelles de la Mecanique Celeste.* Paris, 1899.

[48] J. C. Hall, M. Rosbash, and M. W. Young. https://www.nobelprize.org/prizes/medicine/2017/press-release, 2007.

[49] B. Rossi and D. B. Hall. *The Physical Review*, 59:223, 1941.

[50] R. Tumulka, S. Goldstein, and N. Zanghi. *Physics Review D*, 94:023520, 2016.

[51] S. Augustine, *The Confessions of S. Augustine, translated by E. Bouverie Pusey.* Oxford: John Henry Parker, 1838.

[52] A. Schopenhauer. *The World as Will and Idea, Characterisation of the Will to Live Supplements to the Second Book: Chapter XXVIII,R. B. Haldane and J. Kemp translation. full text: https://heinonline.org;wq: https://heinonline.org.* Kegan, Paul, Trench and TruebnerCo London, 1898.

[53] J. Searle. https://www.youtube.com/watch?v=_rzfstpjgl8.

[54] L. Smolin. *The Trouble With Physics.* Houghton Mifflin, Harcourt, and Penguin (UK), 2006.

[55] L. Smolin. *Studies in History and Philosophy of Science Part B: Studies in History and Philosophy of Modern Physics*, 52, Part A:86–102, 2015.

[56] K. K. Steincke. *Farvel Og Tak ("Goodbye and Thanks").* This attribution is from https://quoteinvestigator.com/2013/10/20/no-predict/, 1948.

[57] M. Tegmark. *Chaos, Solitons & Fractals*, 76:238–270, 2015.

[58] E. Viebahn. *Synthese*, 197:2963–2974, 2020.

[59] G. J. Whitrow. *The Natural Philosophy of Time.* Harper and Row, New York USA, 1963.

[60] G. J. Whitrow. *What is Time?* Oxford University Press USA, 2004.

[61] M. Zukowski and C. Brukner. *J. Phys. A: Math. Theor.*, 47:424009, 2014.

Index

Printed in the United States
by Baker & Taylor Publisher Services